Family of High-Ordered Integer-Valued Auto-Regressive Models and Applications

This book tackles the complexities of integer-valued time series analysis, focusing on over-dispersion, excess zeros, and non-stationarity. It explores high-ordered INAR(p) models with diverse thinning mechanisms and innovation distributions, finding CML superior for inference. Addressing periodicity, harmonic functions are introduced for COVID-19 data. Novel BINAR (1) models with BPWE and SPWE innovations are applied to stock transactions, while new BPGL and SPGL bivariate distributions analyze crime data.

The book derives methodologies, tests performance via simulation, and provides real-life applications, filling a gap in existing literature. This comprehensive work significantly advances the field of integer-valued time series analysis by addressing key challenges such as over-dispersion and periodicity. The detailed exploration of high-ordered INAR(p) models under various thinning mechanisms and innovation distributions provides valuable insights into their performance, with the clear outperformance of the CML inferential method offering practical guidance for researchers. The innovative incorporation of harmonic functions to model the periodic nature of the COVID-19 data in Mauritius demonstrates a crucial adaptation to real-world phenomena. Furthermore, the development and application of novel BINAR (1) models and bivariate distributions like BPGL and SPGL expand the analytical toolkit for understanding the relationships between multiple integer-valued series, exemplified by their application to stock transactions and crime data. By deriving new methodologies, rigorously testing their performance through simulation, and illustrating their utility with diverse real-life applications, this book offers substantial theoretical and practical contributions to the field, addressing limitations in existing literature.

The target audience includes researchers, statisticians, and practitioners working with count data and time series analysis in fields like econometrics, finance, epidemiology, and criminology.

Soobhug Ashwinee Devi works as a Statistician/Senior Statistician at Statistics Mauritius. She is affiliated with the Ministry of Finance, Economic Planning and Development in Port Louis, Mauritius. She is a prominent academic researcher known for her significant contributions to the fields of statistics and public health.

Mamode Khan Naushad is an Associate Professor of Statistics, at the University of Mauritius. His research interests are statistical modeling and computing applied to integer-valued time series modeling.

Sunecher Yuvraj is a Senior Lecturer of Finance and Statistics at the University of Technology Mauritius since 2011. He received his Post Doctorate in Statistical Modeling from the University of Bahia, Ph.D. in Statistics from the University of Mauritius, Master in Business Administration from the Management College of South Africa, Master in Financial Economics and Degree in Mathematics from the University of Mauritius.

Family of High-Ordered Integer-Valued Auto-Regressive Models and Applications

Soobhug Ashwinee Devi, Mamode Khan Naushad,
and Sunecher Yuvraj

CRC Press
Taylor & Francis Group
Boca Raton London New York

CRC Press is an imprint of the
Taylor & Francis Group, an **informa** business
A CHAPMAN & HALL BOOK

First edition published 2026
by CRC Press
2385 NW Executive Center Drive, Suite 320, Boca Raton FL 33431

and by CRC Press
4 Park Square, Milton Park, Abingdon, Oxon, OX14 4RN

CRC Press is an imprint of Taylor & Francis Group, LLC

© 2026 **Soobhug Ashwinee Devi, Mamode Khan Naushad, and Sunecher Yuvraj**

ISBN: 978-1-041-15055-8 (hbk)
ISBN: 978-1-041-15053-4 (pbk)
ISBN: 978-1-003-67745-1 (ebk)

DOI: 10.1201/9781003677451

Typeset in Nimbus font
by KnowledgeWorks Global Ltd.

Contents

Preface

Count or integer-valued time series which consist of discrete quantifiable observations with non-negative structures, occur in diverse physical settings. Examples of count time series are the number of diabetic patients visiting the dialysis center for treatment in successive weeks, the monthly counts of fatal road accidents in Mauritius, and the most recent daily number of new active cases and deaths related to the Novel Coronavirus 2019 (COVID-19) pandemic.

The analysis of integer-valued time series of counts is associated with numerous challenges, since the counting series observations are often influenced by the over-dispersion phenomenon, excess zeros or other integers, time-varying covariate effects, periodicity, and non-stationarity amongst other features. Thus, the integer-valued time series models must be flexible enough to accommodate all these aforementioned features. In the literature so far, the most explored integer-valued autoregressive time series (INAR) model is of order one (INAR(1)). The INAR (1) is composed of, firstly, a survival term based on the thinning principle and an immigrant or innovation term to handle any error effects. In the INAR (1) models so far, there have been a lot of variations in the thinning operators and innovation distributions in an attempt to model the over-dispersion and the other earlier-mentioned features more efficiently. Such variations include the use of Binomial, Generalized Binomial, Negative-Binomial thinnings, and several probability distributions which can be categorized as Generalized Poisson or Poisson mixtures, and were explored to simulate the innovation part. With the rise in the number of applications, it is seen that the INAR (1) becomes limited to model series with high orders, and this requires a high-ordered INAR process, better termed, as an INAR(p) model.

The INAR(p) model has been studied in the literature to some extent, and until recently, the INAR(p) was explored under different thinning algorithms and with Poisson and Negative Binomial (NB) innovations. It is noteworthy that in the recent INAR(p) version, the non-stationarity was accommodated by assuming time-dependent covariates in the innovation predictor functions. However, there are yet no works to confirm the effectiveness of the INAR(p) where the data exhibit excess zeros, or the performance of the INAR(p) with other different innovation distributions, and the extension of such INAR(p) to bivariate INAR(p) models. Within this framework, this book first explores some high-ordered integer-valued time series models of auto-regressive nature (INAR(p)), where the relation between the current and past observations is connected under different thinning mechanisms, including the Binomial, Generalized Binomial, Negative-Binomial and under different innovation distributions that include the popular discrete Generalized Poisson models or mixtures, such as the Conway-Maxwell Poisson (COM-Poisson),

Poisson-Gamma (marginally, the Negative Binomial), Poisson-Lindley (PL), Poisson Inverse-Gaussian (PIG), Poisson Tweedie (PT), Weighted Cosine-Geometric (WCG), Modified-Geometric, Poisson Weighted Exponential (PWE) and the most recent Poisson Generalized Lindley distribution (PGLD) models and their associated zero-inflated versions. The consideration of the different thinnings and innovation distributions is important to assess the performance of the INAR(p) in capturing over-dispersion of various magnitudes. However, execution of Generalized Binomial (GB) and NB thinnings in comparison to binomial thinning under non-stationary and stationary settings is both complex and a very time-consuming procedure and thus should be used restrictively. In terms of the inferential methods, the Conditional Maximum Likelihood (CML) outperformed the Conditional Least Squares (CLS) approaches under all scenarios. Specifically, the performance of the proposed high-ordered INAR processes under CML and CLS methods has been assessed through Monte Carlo simulation experiments. Based on the simulated results, high-ordered INAR processes with Zero Inflated-NB (ZI-NB), ZI-PWE, ZI-PGLD, ZI-Geometric, ZI-WCG, ZI-CMP, and ZI-P innovations under CML provided less biased estimates; however, some computational failures with the CML approach under GB thinning were observed with the ZI versions of CMP, PT, PIG, PL, and WCG innovations. These ZI-models were applied to real-life data sets which in our case, was the COVID-19 new infection case series in Mauritius, which has lately depicted huge volatility patterns, shocks, and consequently non-stationary trends. Here, after a thorough comparative study between the different mentioned ZI innovation distributions and also the thinning operators for the case of Mauritius, an INAR (7) with ZI-NB, ZI-PGLD, and ZI-PWE under the Generalized Binomial thinnings provided the lower AIC, though the computational complexities. The COVID-19 Stringency Index, which measures the degree of strictness of sanitary restrictions, was highly significant in reducing the spread of the virus in Mauritius, followed by other factors like rigorous vaccination campaigns.

As an addition to this book, an important feature of the COVID-19 new infection and death series of Mauritius is periodicity, or the presence of harmonic effects. This interesting yet unexplored aspect allowed the reformulation of the classical INAR process by considering the class of high-ordered INAR models with harmonic innovation, based on the binomial thinning procedure. The latter thinning was more executable in this context and also allowed adherence to the rule of parsimony. This book is an attempt to cater for the periodic feature with some harmonic functions in the definition of the innovation predictor function, and some simulation experiments have been executed to validate the new models and their estimation procedures. The performance of the estimates was found to be less biased. The characteristics of the COVID-19 new infection and death series of South Africa, one of the countries where the most virulent variants have been detected, and that of Mauritius, which is one of the best-performing Sub-Saharan-African small developing states, were found suitable to assess the feasibility of the proposed periodic INAR(p) with selected Poisson mixtures and their zero-inflated and hurdle versions and innovation distributions. Bearing in mind the strong interrelation between the COVID-19 new infection cases and death series, a novel first-order autoregressive bivariate periodic INAR model (periodic BINAR(1)) was also developed whereby the COVID-19 new infection and death cases in Mauritius have been assumed to the ZI-NB, ZI-PWE, and ZI-PGLD innovation distributions because of their huge over-dispersion phenomenon, long sequence of zeros, and non-stationarity in the series. Under the binomial thinning procedure, the periodic BINAR (1) processes with ZI-NB innovation, performed better with comparatively lower AIC, on both COVID-19 series in Mauritius.

Based on real-life scenarios, especially in the financial or social realm, it is known that two series can also be related; for instance, in the financial area, the intraday transactions of particular stocks influence the behavior of other competitive stocks. In this line, since the PWE and PGLD performed satisfactorily on the COVID-19 series, it will be of great interest to explore a novel and flexible bivariate-PWE model using the approaches by Gomez Deniz et al. [2012] and Bermudez and Karlis [2021] based on binomial thinning operators that can account for both positive and negative cross-correlation. The application is on the AXP and GE intraday stock transaction series from the August 1995 New York Stock Exchange, and a comparative study between the proposed BINAR (1) models with BPWE and SPWE distributed innovations was carried out and estimated under the CML and CLS methods. Additionally, some new bivariate distributions based on the Poisson Generalized Lindley marginals were derived. These models include the basic bivariate Poisson Generalized Lindley (BPGL) and the Sarmanov-based bivariate Poisson Generalized Lindley (SPGL) distributions. Subsequently, the BPGL and SPGL are introduced as joint innovation distributions in a novel bivariate first-order autoregressive process (BINAR(1)) based on the binomial thinning procedure. The model parameters in the BPGL and SPGL are estimated using the method of maximum likelihood (ML), while the conditional ML (CML) is used for the BINAR (1) process. The small and large sample performances have been assessed through some Monte Carlo simulation experiments. The new BINAR (1) processes are applied to the criminal records of drug activities (CDRUGS) and shooting activities (CSHOTS) in the 12th police car beat in Pittsburgh for the period from January 1990 to December 2001 and are shown to provide better fitting criteria than other competing BINAR (1) models in the literature.

With the wide range of data applications with complex yet interesting features and also as a means to add value to the existing literature on bivariate models, this work also extends the diagonal BINAR (1) process to the BINAR(p) process based on the common binomial thinning procedure. The proposed BINAR process offers huge flexibility in terms of choice of the innovation distribution to suit the data structure observed in the COVID-19 new infection and death series of Mauritius while accommodating for the specification of explanatory variables such as the age of the patients, the presence of major comorbidities such as cardiovascular diseases, smoking status, and the legislative policies in the country, such as the COVID-19 Stringency Index. The novel BINAR(p) also proved to be flexible enough to accommodate for a wider range of innovation distributions through its computational complexity.

In this book, every intriguing feature of the COVID-19 series was thoroughly studied via high-ordered auto-regressive models with novel Poisson-mixtures and their Zero-Inflated and Hurdle versions as innovation distributions under different thinning operators. Reliable results were obtained and served as the basis for further research, that is, allowed us to shift to the bivariate processes from univariate processes under high-ordered integer-valued time series. Different applications were studied, and under each scenario, the most influential covariates were found. In some cases, the forecasted results can guide policymakers in formulating evidenced-based, comprehensive restorative health and economic measures.

As a future endeavor, given that a close spatial correlation in the COVID-19 series among neighboring countries has been observed, especially with the recent emergence of country-specific variants, it will be an addition to this work to explore the Spatial Integer-Valued Auto-regressive (SINAR) Models with Poisson-mixtures innovation distributions and the most recent spatio-temporal autoregressive (STAR) model for the integer-valued case, which is indexed by space and time simultaneously. Extending these models by having moving average (MA) terms and hence STARMA (or even STARIMA or seasonal STARIMA models) is part of the future work as well.

1

Introduction

Time series of counts are encountered in almost every discipline of research, ranging from the epidemiological [Cardinal et al., 1999, Yu et al., 2013], environmental [Cui and Lund, 2009, Scotto et al., 2014b], meteorological [Tukey, 1949], bio-pharmaceutical [Al Osh, 2009], economics and financial [Blundell et al., 1999, Brännäs and Hellström, 2001, Quoreshi, 2006], and transport [Karlis and Xekalaki, 2005, Sunechar et al., 2018b, Jowaheer et al., 2019, Mamode Khan et al., 2021] to other interdisciplinary studies [Weiß, 2018, Brannas and Quoreshi, 2010, Bourgignon et al., 2018, Bareto Souza, 2015, Lambert, 1992, Hellström, 2001].

Some specific examples in these fields include the intra-day transaction series of stocks [Brannas and Quoreshi, 2010, Kirchner, 2017, Sunechar et al., 2018a, Quoreshi et al., 2020], the daily frequency of claims in insurance [Gouriéroux and Jasiak, 2004, Boudreault et al., 2006, Cossette et al., 2002], the number of transactions per minute for the stock Ericsson B [Joe, 2019], the weekly number of fatal accidents [Karlis et al., 2008, Brijs et al., 2008, Mamode Khan et al., 2021], the monthly number of domestic violence cases [Bakouch and Ristic, 2010], the daily number of antibiotic dispensing medication for the treatment of respiratory diseases registered in a particular health center [Filho et al., 2021], the quarterly number of generic for different medical substances in Sweden [Hellström, 2001], the weekly number of meningococcal disease cases in Germany [Joe, 2019] and Venereal lymphogranuloma (LGV) cases in Barcelona [Morina et al., 2020], the daily number of hotel specific guest nights [Brännäs et al., 2002], the weekly number of sales of a particular product [Böckenholt, 1998, Livio et al., 2018], the monthly number of drug cases in Pittsburgh [Armstrong and Green, 2017] and, the recent daily number of new COVID-19 infection cases and its corresponding number of deaths of a particular state or country [Mamode Khan et al., 2021, Tawiah et al., 2021]. More examples can be found in the *tscount* package in R statistical software [Liboschik et al., 2017].

1.1 Motivating Example

In these mentioned examples, especially in the time series of new COVID-19 infection and death cases for Small Island Developing States (SIDS) such as Mauritius, as illustrated in Figure 1.1, and in South Africa, where the mutation rate of SARS-CoV-2 is comparatively high (refer to Figure 1.3), different data structures with

DOI: 10.1201/9781003677451-1

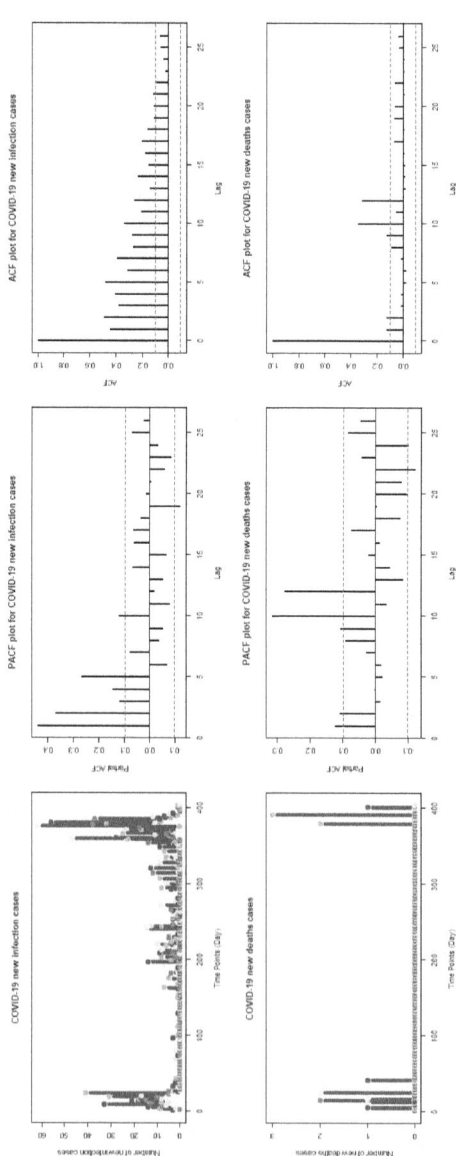

FIGURE 1.1

The time plot, PACF, and ACF of new COVID-19 infection and death cases in Mauritius, 18 March 2020 to 25 April 2021

Descriptive statistics	Tests results	
	New COVID-19 cases	New death cases
Number of Observations	404	404
Mean	2.995	0.040
Variance	57.911	0.068
Jan Van den Broek test for zero-inflation	p-value = 0.000	p-value = 0.000
Over-dispersion test using qcc	p-value = 0.000	p-value = 0.000
Cox-Stuart test for presence of trend	p-value = 0.000	p-value = 0.066
Box-Ljung	p-value = 0.000	p-value = 0.013
Order	7	7

TABLE 1.1
Descriptive statistics and test results for COVID-19 active cases and death series in Mauritius

intriguing characteristics have been observed. Furthermore, the respective data summaries, presented in Table 1.1, statistically demonstrate the effect of over-dispersion which is primarily due to the excess number of zeros in the data or the effect of time-varying and time-independent covariates in Mauritius. Figure 1.1 represents 404 new COVID-19 infection and death case observations each, with respective Partial Auto-Correlation Function (PACF) and Auto-Correlation Function (ACF) plots. The COVID-19 series in Mauritius ranges from the day when the first case was detected, which was on 18 March 2020, until 25 April 2021. Mauritius, as a SIDS located in the Indian Ocean with a population of around 1.3 million and blessed with an economy heavily dependent on the tourism and hospitality sector and international trade, found itself engulfed by the first wave of the COVID-19 pandemic. At the start, an increasing trend in the COVID-19 infection and related death cases was reported, but with prolonged strict sanitary restrictions like mandatory wearing of face masks in public areas, prolonged national lockdown and sanitary curfew, institution of a sufficient number of new quarantine centers, the number of new COVID-19 infections and death cases started to decline. This decreasing trend continued until October 2020, but henceforth, due to the reopening of the Mauritian border, the number of imported COVID-19 infection cases in quarantine centers started to peak. In March 2021, Mauritius was unfortunately swept by a second wave of the COVID-19 pandemic, and until April 2021, the number of new COVID-19 infection cases kept on climbing. The ACF plots for the COVID-19 new infection and death cases, as shown in Figure 1.1, clearly show slow decaying of the spikes, thus confirming the presence of non-stationarity, and the respective PACF plots even confirm that the COVID-19 new infection and death series in Mauritius for that specified period are high-ordered ($p = 7$), qualifying for a high-ordered integer-valued auto-regressive time series structure.

Based on the *qcc.overdispersion.test* in R statistical software via the *'qcc'* package, it is confirmed that the series is over-dispersed (variance > mean), and the Van den Broek test was also significant, proving that the series is zero-inflated. In fact, from June 2020 until October 2020, a long sequence of zero COVID-19 new infection cases was reported. Further, the Ljung-Box test ascertains the existence of serial correlation in the series. The presence of trend via the Cox-Stuart test was also significant. In fact, via the Cox-Stuart test results, it was found that the COVID-19 new cases series has a decreasing trend, whilst the death series does not have an increasing trend. So, the presence of a trend did induce non-stationarity in the series [Brockwell and Davis, 1991, Kim and Park, 2008].

More interestingly, the *fisher.g.test* from the *GeneCycle* package in R and the periodograms below, depict hidden periodicity in the COVID-19 new infection and death series of Mauritius. Based on Figure 1.2, it can be observed that in Periodogram 1, there is a dominant spike at around a frequency of 0.012, so there is a conclusion that there is a dominant periodicity of about 80 days ($1/0.012$). As for Periodogram 2, the duration of periodicity changes, ranging from 2 to 100 days.

Another Sub-Saharan country which caught our attention is South Africa, which, with a slogan like 'Let's grow South Africa together' and blessed with a wide range of distinct ecosystems, is one of the eight African countries bearing the tag of an upper-middle-income country. However, South Africa witnessed a sense of vulnerability in its health system when it recorded the first case of COVID-19 on 5 March 2020 and, as of now, has the highest case incidence on the African continent. The impact of the exponential number of deaths related to COVID-19 following the virulent mutation of SARS-CoV-2 marked South Africa as the National State of Disaster [Sekyere et al., 2017]. The new COVID-19 new infection and deaths series as well as respective PACF and ACF plots are illustrated graphically in Figure 1.3. From Figure 1.3, it can clearly be observed based on the 628 observations of COVID-19 new infection cases, that South Africa witnessed five waves from 05 March 2020 till 22 November 2021, with most noticeable peaks in July 2020, from December 2020 to January 2021, and from June 2021 to August 2021. Despite several sanitary measures imposed and more than 97% of herd immunity achieved, the high death tolls were triggering panic among the South Africans. From the COVID-19 new death series, three peaks were observed in October 2020, from December 2020 to February 2021, and from July to September 2021. During these phases, the BA.4 and BA.5 Omicron emerged, leading to a huge number of hospitalizations and related deaths. In terms of statistical features, the COVID-19 new infection cases and death cases in South Africa demonstrate a huge level of dispersion with a mean of 4665.9 and variance of 27928347 and a mean of 147.8 and variance of 23219.9, respectively. This is also confirmed via the *qcc.overdispersion.test* which provided a p-value, less than the level of significance of 5%. The respective PACF plots also confirm the high orders (orders ≥ 15) and high level of serial autocorrelation. The periodicity test, namely the *ptestg* test from the *ptest* package in R, along with the Cox-Stuart test from the *randtests* package, even confirms the presence of harmonic and non-stationary trends in both series.

Shifting the focus towards developed countries like France and Germany, reference is made to Doukhan et al. [2021]'s works. Here, it is shown that not only SIDS but also the COVID-19 new infection and new death cases in developed countries

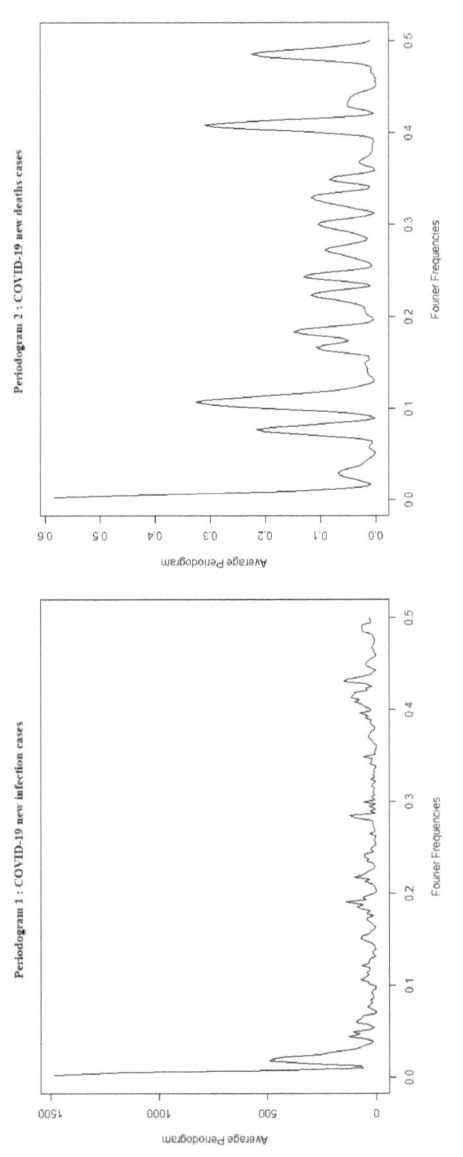

FIGURE 1.2

The periodograms for new COVID-19 infection and death cases in Mauritius

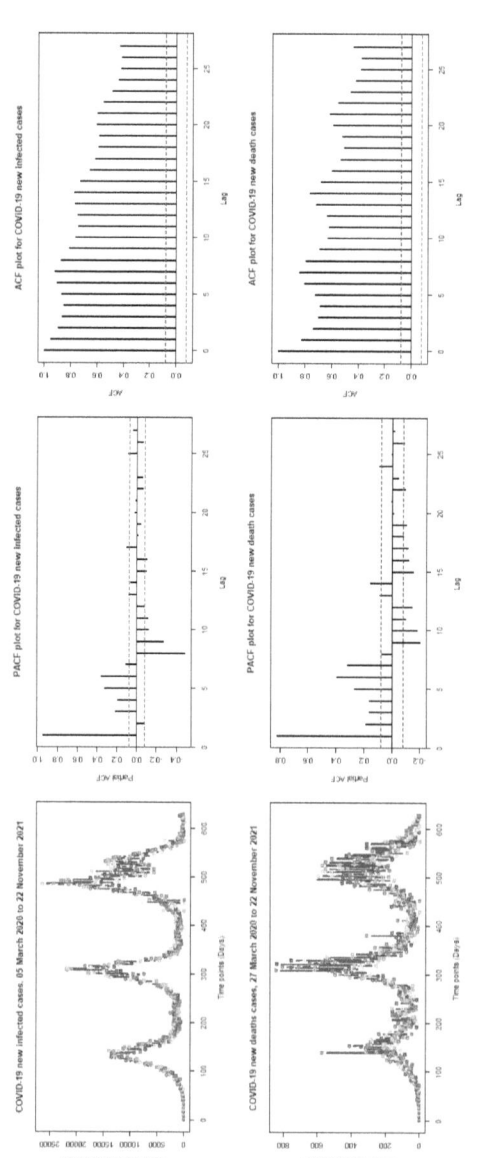

FIGURE 1.3

The time plot, PACF, and ACF of new COVID-19 infection and death cases in South Africa, 05 March 2020 to 22 November 2021 also depict interesting features. Here, the presence of trend, non-stationarity, and, most interestingly, periodicity in both series have also been noted via the Cox-Stuart test and based on the ACF plot, respectively. Based on the above respective PACF plots, both countries' COVID-19 new cases and death series had very high orders and are over-dispersed. Actually, France reported 4 waves of the COVID-19 pandemic notably in April 2020, from November to December 2020, in April 2021, and in July 2021, as compared to Germany, which reported 3 noticeable COVID-19 resurgences in April 2020, from November to December 2020, and in April 2021. So, possibly at these specific periods, different trends can be observed along with some periodic component. It is important to note that these peaks in the COVID-19 cases in each country are reported in the same month each year, possibly due to covariate effects like environmental factors, health-related policies amongst others. France and Germany, based on their respective daily COVID-19 new infection and new death cases ACF plots, as illustrated in Figure 1.4, also have a certain periodic behavior at $S = 7$.

In Figures 1.1 to 1.5, the different features of the new COVID-19 infection and death series have been demonstrated, but interestingly, whether it is in developed or developing countries, the characteristics of both series are quite similar. Some striking features remain the high order of the series, non-stationarity, excess of zeros, and periodicity. Thus, these phenomena help to explain that the time series model suitable to analyze these data must be of high-ordered and flexible enough to accommodate these mentioned features in addition to the over-dispersion. The construction of such models builds the core motivation of this book.

1.2 Rationale and Research Gaps in Current Literature

Until now, in the existing literature, the time series models suited to analyze the time series of counts emanate from the family of Integer-valued Auto-Regressive (INAR) models, a subclass of the family of Integer-valued Auto-Regressive Moving Average (INARMA) models, described in McKenzie [1986, 1988] and Al Osh and Alzaid [1987]. Therein, the time series models in such a class of autoregressive models consist of a survival part and an innovation part [Weiß, 2018, Popovic et al., 2018a]. The survival part is handled by the thinning mechanism [Steutel and Van Harn, 1979, Scotto et al., 2015], while the innovation part is mimicked by a discrete probability model. In fact, the thinning operator connects the current response with the previous-lagged response, and this induces the serial correlation in an observation-driven manner, as compared to parameter-driven models discussed in Cox [1981], Davis and Wu [2009], Neuhaus et al. [2013], Campbell [1994], Brannas and Johansson [1994], Davis et al. [2000], Safari et al. [2011], Bauwens and Veredas [2004], and Hafner and Manner [2012], where correlated random effects are used to induce the serial correlation, or alternatively in Jacobs and Lewis [1978a,b]. Usually, the choice of the innovation distribution is characterized by the features of the data series, such as excess zeros, level of dispersion, covariate effects, and amongst

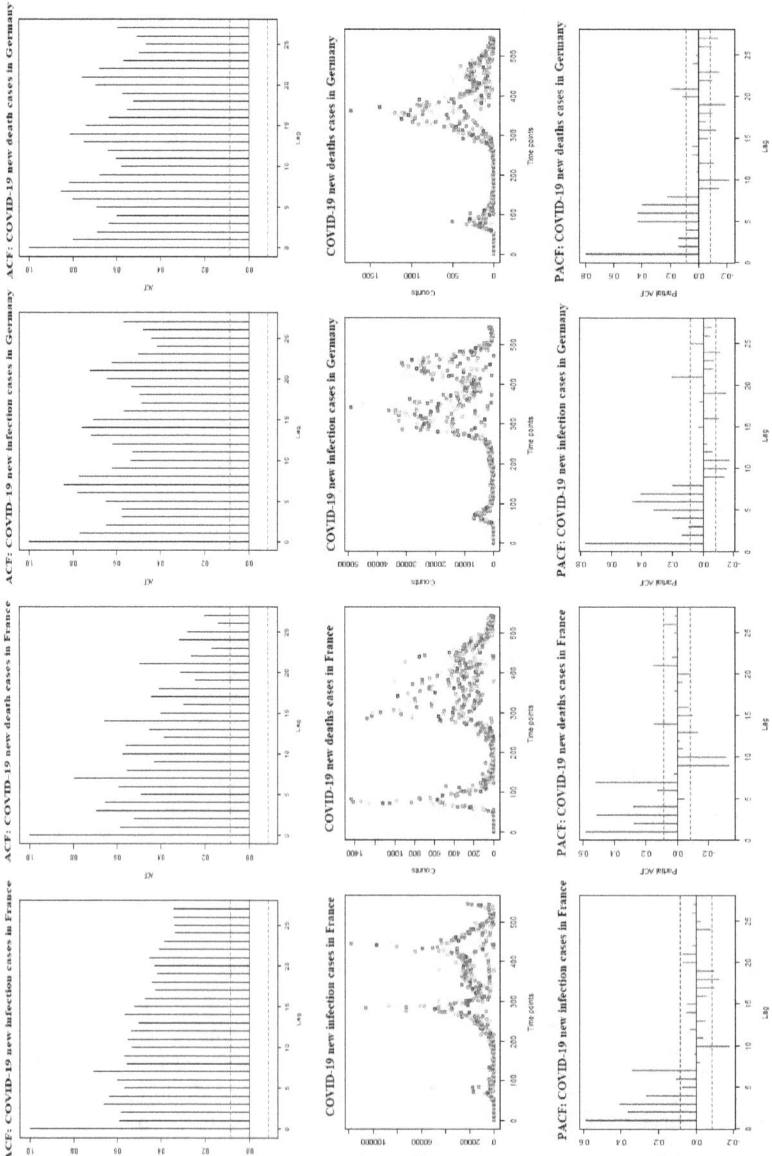

FIGURE 1.4
The ACF, time plots, and PACF of COVID-19 new cases and deaths in France and Germany, 2020 to 2021

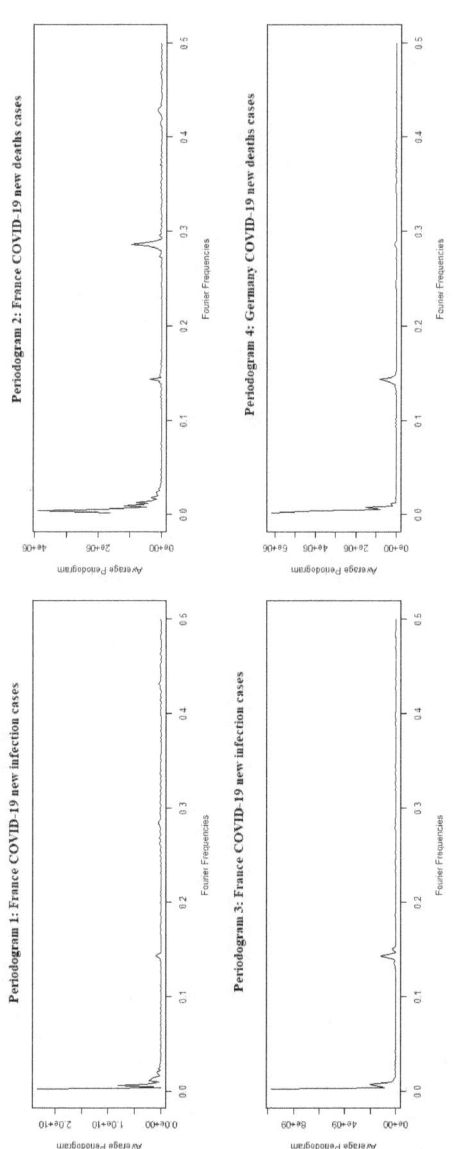

FIGURE 1.5

The periodograms for new COVID-19 infection and deaths cases in France and Germany

others. The thinning operator [Weiß, 2018, Ristic et al., 2009, Awale et al., 2019] and the choices of Poisson-mixture or Generalized Poisson innovation distributions have led to an influx of Integer-valued Auto-Regressive models of order 1 (INAR (1)) [Ristic et al., 2009, Bourguignon and Vasconcellos, 2015, Weiß, 2018, Livio et al., 2018, Sunechar et al., 2018a, Bourgignon et al., 2018] and their bivariate extensions [Pedeli and Karlis, 2011, Scotto et al., 2014b, Ristic et al., 2012, Zhu and Joe, 2010b, Altun, 2019] with zero-inflated innovations [Bakouch and Ristic, 2010, Bareto Souza, 2015, Bourguignon, 2018].

Notably, major attention has been directed to the INAR (1) model structures, while only a few papers considered the high-order INAR(p) processes ($p > 1$) that seem suitable to the data series illustrated in Figures 1.1 and 1.3. The research papers which referred to the INAR(p) processes consist mainly of the works of Alzaid and Al Osh [1990], Du and Li [1991], Bu et al. [2008], Pedeli et al. [2015b], Lu [2018], and the most recent by Joe [2019] and Lu [2019]. The INAR(p) construction is viewed as a direct extension of the single INAR (1) model, but the INAR(p) is associated with some additional properties. Likewise, Alzaid and Al Osh [1990] assumed the paired thinning coefficients to correlate, while Du and Li [1991] relaxed this assumption and proposed a lesser complicated INAR(p) model. Even though the inferential procedures in an INAR(p) process are relatively cumbersome [Bu et al., 2008, Lu, 2018].

Most importantly, the existing INAR(p) do not fully accommodate for the features shown in Figures 1.1 to 1.5. In fact, Joe [2019] provides a very comprehensive exposure of INAR(p) processes under different thinning operators that include the much-used Binomial and Generalized Binomial thinnings, with Poisson and Negative Binomial innovations only. Joe [2019] considered the stationary INAR(p) with respect to the thinning coefficients but allowed time-dependent covariate effects in the innovation distribution specification. The model parameters were estimated using the Conditional Maximum Likelihood function (CML), constructed from the Probability Generating Function (PGF) approach in Davies [1973].

The CML approach by Joe [2019] overcomes the shortcomings in the saddle point approximated likelihood in Pedeli et al. [2015b] and the Taylor approximated likelihood in Lu [2018], and above all, provides more statistically efficient estimates than the Yule Walker (YW) or Method of Moments (MoM) in Du and Li [1991] and Bu et al. [2008]. It is also noted that Zhang et al. [2010] studied the INAR(p) process under strict stationarity conditions and used the Conditional Least Squares (CLS) to estimate the model parameters, but there is yet no comparison between CML and CLS approaches for an INAR(p) process with the non-stationarity features discussed in Figures 1.1 to 1.5 and with time-dependent innovation distributions, especially with regard to the computational performance of these two approaches. Note that, in Bu et al. [2008], the authors compared the CML and the CLS via the Asymptotic Relative Efficiencies (ARE) for the strict stationary INAR(p) process with Poisson innovations, where the AREs were clearly less than one and hence promote the usage of the CML approach. It is worth mentioning that the comparison of YW with approaches such as CML and CLS has already been studied for INAR-type models [Silva et al., 2005, Bourguignon and Vasconcellos, 2015], and it is shown that CML and CLS yield more superior estimates than YW. Besides, in Silva et al. [2005], the comparison between CML, CLS, and YW has been accomplished under a strict stationary INAR(p) process, where CML yielded lesser biased estimates. In light of the above, some research gaps are noticed; hence, this book intends to address these shortcomings.

Given the features in Figures 1.1 to 1.5, and especially in a COVID-19-struck era, it becomes important to explore the INAR(p) processes, considering the zero-inflation or the zero-one inflation [Liu and Zhu, 2021] (referring to the data structures of the COVID-19 new infection series in Sao Tome and Principe, Antigua and Barbuda, Bhutan and Vietnam), the periodicity feature, and non-stationarity. In this endeavour, it is proposed to re-explore the INAR(p) process under different thinning procedures with various popular Poisson-mixtures innovations that include the modified Geometric model, the Poisson-Gamma or marginally the Negative Binomial (NB) model, the Poisson-Lindley (PL) model [Livio et al., 2018, Mohammadpour et al., 2018, Mamode Khan et al., 2019], the Poisson-Tweedie (PT) model Bonat et al. [2017], Kokonendji et al. [2004], Jorgensen and Kokonendji [2014], El Shaarawia et al. [2010] which is a class re-grouping Pólya-Aeppli, the Poisson-Inverse-Gaussian (PIG), Conway-Maxwell Poisson (CMP) [Shmueli et al., 2005, Chatla and Shmueli, 2018], the Poisson-Weighted Exponential [Altun, 2019] (PWE), and the most recent Poisson Generalized Lindley (PGLD) [Ghitany et al., 2008, Gomez-Deniz and Calderin Ojeda, 2011, Abouammoh et al.'s, 2015] which has proven to yield better fitting than contemporary Poisson-mixed models and, finally the last but not the least the trigonometric weighted discrete model like the Weighted Cosine Geometric model [Chesneau et al., 2020]. The choice of Poisson-mixture models was mainly to better understand the behavior of some latent classes, or unobserved groups, on the observed group or overall population. Along with that, it is necessary to assess the performance of the INAR(p) with the aforementioned zero-inflated versions. It is worth constructing these INAR(p) models with different innovation distributions, since the marginal distributions of the counting series may be cumbersome to derive [Mohammadpour et al., 2018, Livio et al., 2018]; and besides, the conditional likelihood via the PGF approach can then be suitably implemented.

In addition to Figures 1.1 and 1.3, some interrelation between the COVID-19 new infection cases and COVID-19-related death series for selected countries, have also been noticed. The cross-correlograms below confirm the existence of some cross-correlation.

In the current literature, in the field of bivariate integer-valued time series models, Pedeli and Karlis [2011], Nastic et al. [2016], Popovic et al. [2018a], Mamode Khan et al. [2019], Quoreshi [2006], and Sunechar et al. [2018b] have proposed suitable bivariate INAR (1) models under both strict stationary and non-stationary setups, with regard to the properties of the corresponding innovation distribution. However, following the INAR(p) process, there is yet no bivariate INAR(p) model. In this work, it is considered to extend the INAR(p) to bivariate INAR(p) with suitable innovation distributions. In this process, the bivariate construction seems interesting since the bivariate model has to be flexible enough to consider the negative correlation as well.

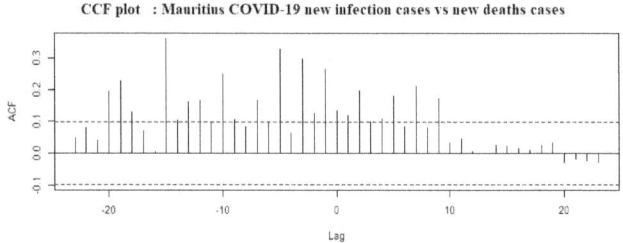

FIGURE 1.6
The cross-correlograms for Mauritius

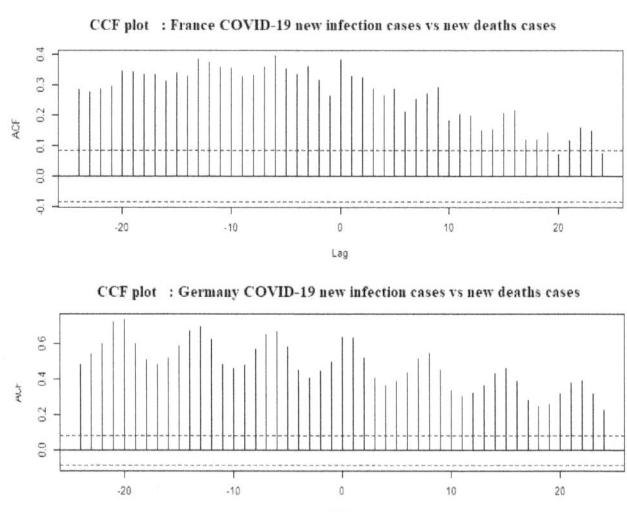

FIGURE 1.7
The cross-correlograms for France and Germany

As for the periodicity, some work in this sense has been accomplished by Bentarzi and Aries [2020], considering that the INAR(p) model does not account for the periodic phenomenon, which was very well present in the COVID-19 new infection case series, demonstrated in the ACF plots in Figure 1.5. This encouraged the study of the so-called periodic time series models, as regarding periodically correlated integer-valued time series, very few contributions are known. In contrast to the continuous case, it is worth emphasizing that the analysis of the periodically correlated series of counts has not received much attention in the literature of count time series. The motivation behind the periodic models is linked to seasonal adjustment. It can happen that somehow seasonally adjusted data still display seasonality. One possible cause may be that the data show periodic properties, and if so, then seasonal adjustment does not make much sense. Periodic data cannot be seasonally adjusted, and the supposedly adjusted series will continue to show signs of seasonality. This holds both in theory and practice, albeit that it is not always immediately obvious and is not easy to detect in real-life data. The implication is that seasonal adjustment, by definition, assumes that a time series can be split up into two independent components, a seasonal and a non-seasonal component. Hence, a periodic model allows the trend and seasonal fluctuations to be related. Moreover, the usual deseasonalization procedure in the Box-Jenkins methodology [Box and Jenkins, 1970] no longer operates in this context of non-negative integer-valued modeling. All these reasons gave us good motivation to extend and generalize the statistical and probabilistic results existing in the time-invariant case to the periodic case.

An immediate application is on the COVID-19 new infection cases time series in a well-known Small Island Developing States (SIDS)—Mauritius, which has depicted huge volatility patterns and shocks, leading to over-dispersion, an excess of zeros, and consequently non-stationary trends. The focus is set on Mauritius since, due to the unprecedented COVID-19 pandemic and given its worldly challenges related to its small size, remoteness, climate resilience, limited economic resources, huge dependence on exportation of either raw materials or partially finished goods, tourism sector as the main economic pillar, huge external debt, heavy reliance on food imports, not-so-favorable balance of trade, high cases of non-communicable diseases (NCDs), and vulnerability to natural calamities, during the peak of the COVID-19 pandemic in 2020, Mauritius witnessed a never-experienced economic contraction of 14.7%. Worse, with the constant depreciation of the Mauritian rupees vis-a-vis other trading foreign currencies and prevailing transport restrictions and logistics bottleneck, huge shortages of certain essential inelastic goods on the domestic market have been observed, and with limited supply comes higher retail prices, which worsen the purchasing power of the Mauritians. Mauritius, backed with strong protection systems but limited economic resources, relies on the effectiveness of the recently imposed fiscal and monetary policies to overcome the economic recession. However, the health of the Mauritian population should not be neglected as well. Based on these arguments, there is the utmost need to timely identify and temper the factors which play a predominant role in the spreading of SARS-CoV-2 in the local Mauritian community.

At the same time, other non-SIDS like South Africa also formed part of this book because their COVID-19 series do depict interesting features. In this book, the proposed novel bivariate models have been applied to different areas of study, namely in financial and social realms. More on these interesting subject matters have been elaborately discussed in Chapters 4 to 6.

1.3   Objectives of the Book

Based on the rationales discussed in the previous sections, the objectives of this book are:

1. Construction of INAR p processes with Poisson mixtures and zero-inflated non-stationary innovation models that incorporate time-varying covariate specification, under the Binomial, Generalized Binomial, and Negative-Binomial thinnings.

2. Investigation of the performance of the Conditional Maximum Likelihood (CML) and Conditional Least Squares (CLS) approaches for the INAR p processes with Poisson mixtures and zero-inflated non-stationary innovation models that incorporate features such as over-dispersion, excess of zeros, covariate specification, non-stationarity, and periodicity. The comparison is to be done via simulations and real-life experiments.

3. Major application areas: The COVID-19 new infection and new death cases time series and situational assessments of the COVID-19 pandemic in Mauritius and South Africa.

4. Explore the novel best-performing Poisson-mixtures innovation distributions and come up with their associated BINAR (1) processes that can account for the harmonic effects, non-stationarity, over-dispersion, and other features demonstrated in Figures 1.3 and 1.5. Application will be on the financial stock markets' data and the Pittsburgh monthly crime data.

5. Extension to Bivariate INAR(p) models with specific innovation distributions.

1.4   Organization of the Book

Chapter 1 outlines the core motivation of this book and highlights the research gaps in existing literature, exemplified by the most recent and motivating real-life data—the COVID-19 new infection and death series for Mauritius. The different features of the latter series have been elaborately discussed and the way forward to address the research gaps has been specified.

Chapter 2 provides an in-depth review of the existing integer-valued autoregressive (INAR) models and their extensions to BINAR processes. The highlight is on the different thinning mechanisms and the choice of the innovation distributions as a means to handle the over-dispersion as well as other features of the series like periodicity, zero-inflation, non-stationarity in the series, and the covariate specifications. Asymptotic properties of the INAR(p) process under a non-stationarity setting have been derived. Further, different estimation procedures, namely the Conditional Least Squares and Conditional Maximum likelihood approaches have been reviewed and derived accordingly.

Chapter 3 focuses on running Monte Carlo simulations. An INAR (4) process under the Binomial and Generalized Binomial thinning for different Poisson-mixture innovation distributions such as the Conway-Maxwell Poisson (COM-Poisson),

Poisson-Gamma (marginally, the Negative Binomial), Poisson-Lindley, Poisson Inverse-Gaussian, Poisson Tweedie, Weighted Cosine-Geometric models, Poisson Weighted Exponential (PWE), and the recently emerged Poisson Generalized Lindley (PGL) and their associated Zero-Inflated versions, assuming stationarity and non-stationarity, have been simulated in R statistical software. Note that the simulations were conducted under the CML and CLS estimation procedures. Based on the simulation results, some good insights on the workability of the different INAR processes can be retained. For instance, the computational failures under GB thinning and of certain models were noted. Conversely, the smooth execution of the binomial thinning under any scenario was observed. All these findings served as a basis when embarking on our next researches elaborated on subsequent chapters.

Chapter 4 focuses on the applicability of the most suitable models on the COVID-19 new infection cases in Mauritius. Different time-varying covariates, namely the reproduction rate (ReR), the COVID-19 risk due to weather conditions (CRW), the major event of vaccination, and the COVID-19 Stringency Index, have been induced in the model to identify the most influential factor on the spread of SARS-CoV-2 in Mauritius. The models with the lowest AIC have been chosen, and the results were interpreted and discussed.

In Figures 1.3 and 1.4, the periodic or harmonic feature in the COVID-19 new infection and death cases was clearly depicted. In chapter 5, we propose some alternative formulations of the classical INAR process by considering the class of high-ordered INAR models with harmonic NB, PWE, and PGLD and their associated zero-inflated and hurdle innovation distributions under the simple binomial thinning operator. By doing so, the level of over-dispersion and excess zeros is being accounted for. The focus is on the COVID-19 new infection and death series of Mauritius and South Africa mainly because Mauritius was the only SIDS to have curbed the propagation of SARS-CoV-2 in 2020 via strict sanitary restrictions, and South Africa was amongst those countries to have detected the most virulent variant—Omicron. In the same line, the periodic INAR(p) has also been extended to the bivariate periodic INAR model.

In Chapter 6, the scope of the study has further been explored, and we proposed a new bivariate INAR (1) model with paired Poisson-Weighted Exponential distribution, applied to the AXP and GEstock transactions data. The simulation results and findings have been discussed extensively.

Under chapter 7, based on simulation results but bounded by the computational complexities when running the high-ordered ZI models, we came up with two novel bivariate distributions, namely, the Basic bivariate Poisson generalized Lindley (BPGL) and Sarmanov bivariate Poisson generalized Lindley (SPGL). Their respective mathematical properties have been derived. The area of interest was the Pittsburgh crime data set, and the application results reveal that BINAR (1)SPGL provides better model adequacy measures than some other recently proposed BINAR (1) models.

After exploring the bivariate INAR (1), an attempt has also been made to extend it to BINAR(p) under binomial thinning and most importantly under a non-stationary setting. In this Chapter 8, the COVID-19 new infection and death cases were dually analyzed while accommodating for explanatory variables. The use of these BINAR(p) models, however, depends on the nature of the data.

Finally, in Chapter 9, the future works have been highlighted with a main focus on exploring the Spatial Integer-Valued Auto-regressive (SINAR) Models with Poisson-mixtures innovation distributions and the most recent spatio-temporal autoregressive (STAR) model for the integer-valued case, which is indexed by space and time simultaneously. An attempt to extend these models by having moving average (MA) terms and hence STARMA (or even STARIMA or seasonal STARIMA models) is also envisaged.

2

State of Art

2.1 The Simple Integer-Valued Auto-Regressive Model (INAR (1))

From McKenzie [1986], the INAR (1) is expressed as

$$Y_t = \rho * Y_{t-1} + R_t \tag{2.1}$$

where $*$ is the thinning operator, in particular the binomial thinning operator [Steutel and Van Harn, 1979], that connects the current response Y_t, at the tth time point, $t = 1, 2, 3, \ldots, T$, with the previous-lagged observation Y_{t-1}. The function of the $*$ operator is to ensure the sequence of $\{Y_t\}_{t=1}^{T}$ has a discrete structure, as compared to the Gaussian time series where the multiplication operator is used instead of the thinning operation [Scotto et al., 2015]. In many literatures, the part $\rho * Y_{t-1}$ is termed as the survival part as well, as described in Weiß [2018] and Popovic et al. [2018a]. The ρ in Equation 2.1 is treated as a fixed non-negative constant that lies in the interval $[0, 1)$ and in particular, this range indicates the stationarity stable condition of the INAR (1) process. On the other hand, $\rho = 1$ usually indicates the non-stationarity condition [Du and Li, 1991, Dion et al., 1995]. In the same context, Kim and Park [2008] and Zhang et al. [2010] assumed another range of ρ, lying in the interval $(-1, 1)$, as the authors were considering differenced series that emanate originally from non-stationary series that assume both positive and negative values. A quite similar treatment on non-stationary series was considered by Andersson and Dimitris [2014], with Skellam distributed innovations. Furthermore, in the area of non-stationary discrete time series, Nastic et al. [2017] induced non-stationarity in their INAR process by assuming different environmental states that were handled by K-means clustering [Weiß, 2018, Moller et al., 2018, Tang and Wang, 2014], but in practice, especially in the SARS-CoV-2 series, quite a big dispersion may be noticed in subsequences with large and frequent ups and downs, and this disables the usage of the K-means clustering, and subsequently the environmental states specification may not serve the purpose.

In this book, the non-stationarity in the data series is handled in a simpler way by assuming time-dependent covariate terms in the innovation $\{R_t\}_{t=1}^{T}$ specification [Weiß, 2015, Jowaheer and Sutradhar, 2002, Sunechar et al., 2018b, Jowaheer et al., 2018, Sunechar et al., 2018a, Schweer and Weiß, 2014, Joe, 2019]. The sequence of $\{R_t\}$ is assumed to be independently distributed with mean μ_t and variance, σ_t^2,

DOI: 10.1201/9781003677451-2

and we also assumed another parameter v that denotes any extra-Poisson dispersion in the discrete r.v, R_t. Under these assumptions in R_t, Weiß [2015] confirms that the INAR process is Markovian. An additional assumption is that the covariance between the previous lagged observations, Y_{t-k}, for $k \geq 1$ and the current innovation term, R_t, that is, $Cov(Y_{t-k}, R_t) = 0$, and hence there exists a covariance between Y_t and R_t, given by $Cov(Y_t, R_t) = V(R_t)$. Based on this assumption of independence between the sequence of $\{R_t\}$ and $\{Y_{t-k}\}$, the conditional probability generating function of $Y_t \mid Y_{t-1}$, denoted by $G_{Y_t|Y_{t-1}}(s) = E[s^{Y_t|Y_{t-1}}]$ is given as:

$$G_{Y_t|Y_{t-1}}(s) = G_{\rho * Y_t|Y_{t-1}}(s) \times G_{R_t}(s), \tag{2.2}$$

where, for, $\rho * Y_{t-1} = \sum_{j=1}^{Y_{t-1}} B_j$, with B_j is identically and independently Bernoulli distributed r.v with fixed probability ρ, that is $\rho * Y_{t-1} \mid Y_{t-1} \sim \text{Binomial}(Y_{t-1}, \rho)$, the Equation 2.2 becomes

$$G_{Y_t|Y_{t-1}}(s) = (1 - \rho + \rho s)^{Y_{t-1}} \times G_{R_t}(s). \tag{2.3}$$

Definitely, by altering the probability properties of $\rho * Y_{t-1}$ [Latour, 1997, 1998; Gauthier and Latour, 1994; Weiß, 2018], different $G_{Y_t|Y_{t-1}}(s)$ expressions are obtained, and hence this leads to different conditional likelihoods. In this book, further to the research by Ristic et al. [2009] and Joe [2019], it is assumed that for fixed ρ, $G_{\rho * Y_{t-1}|Y_{t-1}}(s)$ for the different popular thinnings is expressed as:

- Binomial thinning,
$$G_{\rho * Y_{t-1}|Y_{t-1}}(s) = (1 - \rho + \rho s)^{Y_{t-1}} \tag{2.4}$$

- Generalized Binomial thinning [Joe, 2019, Aly and Bouzar, 1994, 2019],

$$G_{\rho * Y_{t-1}|Y_{t-1}}(s) = \left[\frac{(1-\rho) + (\rho - \gamma)s}{(1 - \rho\gamma) - (1-\rho)\gamma s} \right]^{Y_{t-1}}, \tag{2.5}$$

- Negative Binomial thinning [Ristic et al., 2009].

$$G_{\rho * Y_{t-1}|Y_{t-1}}(s) = \left(\frac{\rho}{1 - (1-\rho)s} \right)^{Y_{t-1}} \tag{2.6}$$

Besides, with the above expressions, the estimation of the parameters become less cumbersome under different inferential procedures and this point is to be further discussed in the subsequent sections and chapters. Further to this, the simulation experiments in Ristic et al. [2009] and Joe [2019] proved that the resulting model estimators are less biased than under the other competing thinning which assume random coefficients. In fact, these different thinning impact on the properties of $\{Y_t\}$ and subsequently become a source to account for the frequent over-dispersion phenomenon noticed in the repeated count measures [Weiß, 2008, Zhu and Joe, 2006, McKenzie, 1986, Turkman et al., 2014, Nastic et al., 2016, Nastic and Bakouch,

2012, Joe, 2019, Bakouch et al., 2017, Awale et al., 2021]. The latter point on handling over-dispersion is further elaborated in the next section of Chapter 2.

The other core component of the INAR process is the sequence of the innovation term $\{R_t\}_{t=1}^T$. The probability distribution of R_t is deemed to be primordial with regard to the data structures. Likewise, in Figures 1.1 to 1.4 in Chapter 1, the main features that drew our attention are the non-stationarity, over-dispersion, excess of zeros, and periodicity. As for the first three mentioned features, these can be handled by the properties of $\{R_t\}$. In the context of handling over-dispersion in INAR models, Al Osh and Aly [1992] and Alzaid and Al Osh [1993] assumed the innovation to be geometric [Ristic et al., 2009, Nastic et al., 2017, Huang and Zhu, 2021] and Generalized-Poisson [Zhu, 2012b, Rasaki, 2018, Ye et al., 2011]. Other over-dispersed INAR (1) models have also been constructed with Negative Binomial (NB) and COM-Poisson [Weiß, 2015, Jowaheer and Sutradhar, 2002, Sunechar et al., 2018b, Jowaheer et al., 2018, Sunechar et al., 2018a, Schweer and Weiß, 2014]. On the other hand, to account for over-dispersion via zero-inflated (ZI) structures, Ristic et al. [2009], Jazi et al. [2012], Bareto Souza [2015], Bourguignon [2018] constructed the INAR (1) with zero-inflated innovation distribution. In these papers, the INAR (1)s with the different innovations have shown to provide reliable fitting with fairly consistent estimates of the model parameters [Weiß, 2015]. However, in the construction of the INAR (1) process, it seems less cumbersome to assume a probability model for the innovation series without compromising on the marginal distribution of the counting series, as the commonest inferential procedures such as CML and CLS estimation methods are formulated using conditional information on the previous lagged observations. Besides, INAR (1) constructed under such perspective, that is, with different innovation distributions have proven to provide better fitting criteria than the counter INAR (1)s (See the papers by Mohammadpour et al. [2018], Livio et al. [2018]). In this sense, the parametrization of the innovation distribution may allow for time-variant covariate specification, which ultimately induces the non-stationary moments, as shown in Weiß [2015] and Joe [2019].

In addition, from Figures 1.6 and 1.7, it is remarked that the INAR series of two measurements emanate from the same variable may be inter-related, like the series of newly infected and deaths due to COVID-19. In this context, it is worth to explore the bivariate INAR (BINAR) model in this book. In the literature, initially, Pedeli and Karlis [2011, 2013b,c] proposed a constrained BINAR (1) model from extending the classical INAR process in McKenzie [1986]. The constrained BINAR (1) is given by:

$$Y_{t,1} = \rho_{11} * Y_{t-1,1} + R_{t,1} \tag{2.7}$$

$$Y_{t,2} = \rho_{22} * Y_{t-1,2} + R_{t,2} \tag{2.8}$$

where $.^{[m]}$ indexes the $m = 1, 2$ variate, and in the above, the same assumptions of the thinning parameters ρ_{11}, ρ_{22} on $R_{t,1}$ and $R_{t,2}$ respectively, as in Equation 2.1 hold. Also, $Cov(Y_{t-j,i}, R_{t,i}) = 0$, $Cov(Y_{t-j,1}, R_{t,2}) = Cov(Y_{t-j,2}, R_{t,1}) = 0$ for $j \in Z^+$ and $i \in \{1, 2\}$. In addition, the inter-relation between the two series $\{Y_{t,1}, Y_{t,2}\}$ is induced by the cross-correlated innovations $\{R_{t,1}, R_{t,2}\}$ where the latter can follow some bivariate distributions under both stationary or non-stationary innovation distribution specification such as the bivariate Poisson [Kocherlakota and Kocherlakota, 2001, Mamode Khan et al., 2016], bivariate NB [Marshall and Olkin, 1990, Ng et al., 2010, Sunechar et al., 2018b, Mamode Khan et al., 2018], bivariate COM-Poisson [Sellers et al., 2016, Jowaheer et al., 2018] or bivariate distributions under copula constructors as in Karlis and Pedeli [2013d], Pedeli and Karlis [2013b] such that $Cov(Y_{t,1}, Y_{t,2}) = Cov(\rho_{11} * Y_{t-1,1}, \rho_{22} * Y_{t-1,2}) + Cov(R_{t,1}, R_{t,2})$ Up to now, the constrained BINAR (1) has been developed under the binomial thinning assumption, and from Al Osh and Alzaid [1987], the individual counting series $Y_{t,i}$ can be expressed as: $Y_{t,i} \overset{d}{=} \sum_{j=0}^{\infty} \rho_{ji,j} * R_{t-j,i}$ and this allows to write the PGF as $G_{Y_{t,i}}(s) = \prod_{j=0}^{\infty} G_{R_{t-j,i}}(1 - \rho_{ji,j} + \rho_{ji,j}s)$ and hence, $G_{\tilde{Y}_t}(s) = G_{Y_{t,1}, Y_{t,2}}(s) = \prod_{j=0}^{\infty} G_{R_{t,1}, R_{t,2}}\left((1 - \rho_{11,j} + \rho_{11,j}s_1), (1 - \rho_{22,j} + \rho_{22,j}s_2) \right)$. As for the inferential procedures in the context of the above BINAR (1), Pedeli and Karlis [2011] proposed the Yule Walker moment based and the CML approach while Mamode Khan et al. [2016] adopted the quasi-likelihood procedure. The estimation procedures are discussed later in this chapter.

In addition, the unconstrained BINAR (1) studied by Ristic et al. [2012], Pedeli and Karlis [2013aa], Nastic et al. [2016], Popovic et al. [2018b] and recently by Bakouch et al. [2021] provide two sources of cross correlation, that may originate from the cross-correlated innovation and or from the previous lagged observations of the counter series, that is, the BINAR (1) is expressed as:

$$Y_{t,1} = \rho_{11} * Y_{t-1,1} + \rho_{12} * Y_{t-1,2} + R_{t,1} \tag{2.9}$$

$$Y_{t,2} = \rho_{21} * Y_{t-1,1} + \rho_{22} * Y_{t-1,2} + R_{t,2}, \tag{2.10}$$

where the same assumptions on the counting series, thinning parts, and innovation terms hold. In fact, in the unconstrained version, Pedeli and Karlis [2013aa] imposes the two sources of cross correlation, and thus makes the model more generic. Refer to Equations 2.9 and 2.10,

$$\begin{aligned}
Cov(Y_{t,1}, Y_{t,2}) = {} & Cov(\rho_{11} * Y_{t-1,1}, \rho_{22} * Y_{t-1,2}) \\
& + Cov(\rho_{12} * Y_{t-1,2}, \rho_{21} * Y_{t-1,1}) \\
& + Cov(\rho_{11} * Y_{t-1,1}, \rho_{21} * Y_{t-1,1}) \\
& + Cov(\rho_{12} * Y_{t-1,2}, \rho_{22} * Y_{t-1,2}) \\
& + Cov(R_{t,1}, R_{t,2})
\end{aligned}$$

with $G_{\tilde{Y}_t}(s) = \prod_{j=0}^{\infty} G_{R_{t,1}, R_{t,2}}\left((1 - \alpha_j + \alpha_j s_1)(1 - \gamma_j + \gamma_j s_1), (1 - \beta_j^* + \beta_j^* s_2)(1 - \delta_j + \delta_j s_2) \right)$ where, $\begin{bmatrix} \alpha_j & \gamma_j \\ \beta_j^* & \delta_j \end{bmatrix} = \begin{bmatrix} \rho_{11} & \rho_{12} \\ \rho_{21} & \rho_{22} \end{bmatrix}^j$.

2.2   The INAR(p) Process

From Equation 2.1, the INAR(p) process is obtained by extending the previous lagged observations to obtain

$$Y_t = \rho_1 * Y_{t-1} + \rho_2 * Y_{t-2} + \cdots + \rho_p * Y_{t-p} + R_t \qquad (2.11)$$

with the same assumptions of the previous lagged terms, Y_{t-k}, $1 \leq k \leq p$ on R_t, that is, $Cov(Y_{t-k}, R_t) = 0$. The $*$ represents the thinning operator with $\rho_k \in [0, 1)$ and we may consider the thinning distribution as listed in Section 2.1. In fact, Joe [2019] worked with the Binomial and Generalized Binomial thinnings. Alzaid and Al Osh [1993] assumed thinning coefficient in $\rho_j * Y_{t-j}$ and $\rho_{j'} * Y_{t-j'}$ to be dependent, that is, $\rho_j * Y_{t-j} = \sum_{m=1}^{Y_{t-j}} b_{jm}(\rho_j), \rho_{j'} * Y_{t-j'} = \sum_{m'=1}^{Y_{t-j'}} b_{j'_{m'}}(\rho_{j'})$, the Bernoulli sequence $\{b_{jm}(\rho_j)\}$ and $\{b_{j'_{m'}}(\rho_{j'})\}$ are dependent, while Du and Li [1991] relaxed this assumption and assumed independence between $\{b_{jm}(\rho_j)\}$ and $\{b_{j'_{m'}}(\rho_{j'})\}$ and hence, from their Lemma 2.1 in Du and Li [1991], $E\{(\rho_j * Y_{t-j})(\rho_{j'} * Y_{t-j'})\} = \rho_j \rho_{j'} E(Y_{t-j} Y_{t-j'})$, and hence, $Cov(\rho_j * Y_{t-j}, \rho_{j'} * Y_{t-j'}) = \rho_j \rho_{j'} Cov(Y_{t-j}, Y_{t-j'})$. Zhang et al. [2010] established the sufficient condition for strict stationarity and ergodicity of the INAR(p) process with the stable conditions $\sum_{k=1}^{p} \rho_k < 1$. Further to the specification on R_t defined in the previous subsection, it is assumed, in this book that $\sum_{k=1}^{p} \rho_k < 1$ and let R_t follow a probability distribution D, that is, $R_t \sim D(Y_{t,1}, Y_{t,2}, \ldots, Y_{t,s}, v)$, with its link or mean function given by:

$$\log \theta_t = \beta_0 + \beta_1 Y_{t,1} + \beta_2 Y_{t,2} + \cdots + \beta_s Y_{t,s} \qquad (2.12)$$

and v is the extra-Poisson or over-dispersion parameter. The vector $\tilde{Y}_t = (Y_{t,1}, Y_{t,2}, \ldots, Y_{t,s})'$ represents the coefficient for the s covariates. Thus, with this specification on R_t, the expectation and variance of Y_t, that is, $E(Y_t)$ and $V(Y_t)$ vary with time, and this becomes a way to depict the trend and non-stationarity in the series as illustrated in Figures 1.1, 1.3, and 1.4 in Chapter 1.

2.3   Fisher Index of Dispersion under the Different Thinnings

The Fisher Index of dispersion under the different thinnings can be computed as:

1. Under the Binomial thinning assumption,

$$\rho * Y_t \mid Y_t \sim Bin(Y_t, \rho)$$

$$E(Y_t) = \sum_{k=1}^{p} \rho_k E(Y_{t-k}) + E(R_t)$$

where $E(R_t) = \mu_t$ and,

$$Var(Y_t) = E(Y_t) + \sum_{k=1}^{p} \rho_k^2 [Var(Y_{t-k}) - E(Y_{t-k})] +$$

$$2 \sum_{k=1}^{p} \sum_{k<k'}^{p} \rho_k \rho_{k'} Cov(Y_{t-k}, Y_{t-k'}) + [Var(R_t) - E(R_t)]$$

The Fisher index of dispersion under the Binomial thinning $(FI_{t,B})$ at time t, is given by:

$$FI_{t,B} = \frac{Var(Y_t)}{E(Y_t)}$$

$$= 1 + \left[\frac{\begin{array}{l} \sum_{k=1}^{p} \rho_k^2 [Var(Y_{t-k}) - E(Y_{t-k})] \\ +2\sum_{k=1}^{p} \sum_{k<k'}^{p} \rho_k \rho_{k'} Cov(Y_{t-k}, Y_{t-k'}) \\ +[Var(R_t) - E(R_t)] \end{array}}{E(Y_t)} \right]$$

Thus, for $\forall\, k \in \{1,p\}$, given $FI_{t-k,B} > 1$ and positive covariances, then $FI_t > 1$, iff, R_t is equi- or over-dispersed.

2. Under the Generalized Binomial thinning, with FI indicated as $FI_{t,G}$

$$E(\rho * Y_t \mid Y_t) = \rho Y_t$$

$$V(\rho * Y_t \mid Y_t) = \rho(1 - \rho)\frac{(1+\gamma)}{(1-\gamma)}Y_t$$

For $\gamma = 0$, we obtain the same properties as the Binomial thinning. It can be shown that,

$$E(Y_t) = \sum_{k=1}^{p} \rho_k E(Y_{t-k}) + E(R_t)$$

$$V(Y_t) = \sum_{k=1}^{p} \left(1 + \frac{2\gamma}{(1-\gamma)}\right)\rho_k(1 - \rho_k)E(Y_{t-k})$$

$$\sum_{k=1}^{p} \rho_k^2 Var(Y_{t-k}) + Var(R_t) +$$

$$2 \sum_{k=1}^{p} \sum_{k<k'}^{p} \rho_k \rho_{k'} Cov(Y_{t-k}, Y_{t-k'})$$

By letting $V(Y_t)$ as $V_B(Y_t)$, then,

$$V(Y_t) = \sum_{k=1}^{p} \frac{2\gamma}{(1-\gamma)} \rho_k (1 - \rho_k) E(Y_{t-k}) + V_B(Y_t)$$

Therefore,

$$FI_{t,G} = \frac{\sum_{k=1}^{p} \frac{2\gamma}{(1-\gamma)} \rho_k (1 - \rho_k) E(Y_{t-k})}{E(Y_t)}$$
$$+ \frac{V_B(Y_t)}{E(Y_t)},$$

where, $\frac{V_B(Y_t)}{E(Y_t)} = FI_{t,B}$ and hence, $FI_{t,G} > FI_{t,B}$. Thus, with the Generalized Binomial thinning, the INAR (p) is capable of modeling high or severe over-dispersion, than the Binomial thinning.

3. For the Negative Binomial thinning, with FI denoted as $FI_{t,N}$,

$$\rho * Y_t \mid Y_t = \sum_{j=0}^{Y_t} B_j(\rho),$$

where, $B_j(\rho)$ is i.i.d sequence of Geometric r.v with probabilities, ρ, that is $f(j;\rho) = (1-\rho)^j \rho$, and hence, from the properties of Negative Binomial,

$$E(\rho * Y_t \mid Y_t) = \frac{(1-\rho)}{\rho} Y_t,$$

$$Var(\rho * Y_t \mid Y_t) = \frac{(1-\rho)}{\rho^2} Y_t$$

and for the INAR(p),

$$E(Y_t) = \sum_{k=1}^{p} \left(\frac{1-\rho_k}{\rho_k} \right) E(Y_{t-k}) + E(R_t)$$

$$V(Y_t) = \sum_{k=1}^{p} \left(\frac{1-\rho_k}{\rho_k} \right) E(Y_{t-k})$$
$$+ \sum_{k=1}^{p} \left(\frac{1-\rho_k}{\rho_k} \right)^2 Var(Y_{t-k}) + Var(R_t)$$
$$+ 2 \sum_{k=1}^{p} \sum_{k<k'}^{p} \left(\frac{1-\rho_k}{\rho_k} \right) \left(\frac{1-\rho_{k'}}{\rho_{k'}} \right) Cov(Y_{t-k}, Y_{t-k'})$$

By letting $\alpha_k^* = \frac{1-\rho_k}{\rho_k}$, we obtain,

$$E(Y_t) = \sum_{k=1}^{p} \alpha_k^* E(Y_{t-k}) + E(R_t),$$

$$V(Y_t) = \sum_{k=1}^{p} \alpha_k^*(\alpha_k^* + 1)E(Y_{t-k})$$
$$+ \sum_{k=1}^{p} \alpha_k^{*2} Var(Y_{t-k}) + Var(R_t)$$
$$+ 2\sum_{k=1}^{p}\sum_{k<k'}^{p} \alpha_k^* \alpha_{k'}^* Cov(Y_{t-k}, Y_{t-k'})$$

$$V(Y_t) = E(Y_t) + \sum_{k=1}^{p} \alpha_k^{*2} E(Y_{t-k}) + \alpha_k^* Var(Y_{t-k})$$
$$+ 2\sum_{k=1}^{p}\sum_{k<k'}^{p} \alpha_k^* \alpha_{k'}^* Cov(Y_{t-k}, Y_{t-k'}) + [Var(R_t) - E(R_t)]$$

and,

$$FI_{t,N} = \frac{Var(Y_t)}{E(Y_t)} = 1+$$

$$\left[\frac{\sum_{k=1}^{p} \alpha_k^* E(Y_{t-k}) + \alpha_k^{*2} Var(Y_{t-k}) + 2\sum_{k=1}^{p}\sum_{k<k'}^{p} \alpha_k^* \alpha_{k'}^* Cov(Y_{t-k}, Y_{t-k'}) + [Var(R_t) - E(R_t)]}{E(R_t)} \right]$$

Hence, $FI_{t,N} > 1$, iff, R_t, is equi- or over- dispersed given additionally the $Cov(Y_{t-k}, Y_{t-k'}) > 0$. Overall, we can see that, under either thinnings, the INAR(p) process is capable of capturing any forms of over-dispersion with over-dispersed innovations.

The estimation of the parameters under the INAR(p) process is quite a challenging procedure. This part is discussed thoroughly in the Section 2.5.

2.4 Inferential Procedures for the INAR and Bivariate INAR Processes

Several attempts have been made to estimate the parameters in the INAR(p) process. To begin with, it is remarked in the INAR-related literature, especially for the INAR (1) and INAR(p) processes under strict stationarity, that the commonest approaches to estimate the model parameters are mainly Yule-Walker approach (YW) [Du and Li, 1991, Silva et al., 2005], the CLS [Klimko and Nelson, 1978, Tjostheim, 1986, Karlsen and Tjostheim, 1988, Livio et al., 2018, Bu et al., 2008], and the CML approach [Bu et al., 2008, Freeland and McCabe, 2004, Bu et al., 2008]. Within these approaches, Bourguignon [2018] and Silva et al. [2005] showed that the CML and CLS provides more superior and less biased estimates than YW, though YW is computationally less cumbersome. In comparison to CML and CLS approaches, for the strict stationary INAR(p) process, Bu et al. [2008] and Silva et al. [2005] illustrated through simulation experiments that CML yields lesser biased estimates. This prompts other researchers like Pedeli et al. [2015b], Lu [2019] and Joe [2019] to further explore the INAR(p) with different innovation specifications. However, the CML approach reports quite a number of computational issues. Likewise, Pedeli et al. [2015b] proposed the saddle point approximation to the CML approach, but as Lu [2019] reported that this approximation lacks closed-form expressions, and the computation demands numerical inversion of nonlinear functions, and this results in a significant amount of approximation errors and hence inconsistent simulated mean estimates. Joe [2019], however, reported that the alternative numerical technique by Lu [2019] using a Taylor approximated likelihood does not adapt to the different thinning other than Binomial thinning.

In this context, in this book, we pursue with the CML computational procedures proposed by Joe [2019], that are based on the inversion of the PGF of the conditional likelihood of the INAR(p) process to obtain the likelihood function [Davies, 1973, Zhu and Joe, 2010a,b]. However, since both the CML and CLS approaches have, so far, not been explored for the INAR(p) process with the non-stationary innovations that can be zero or non-zero inflated, it is proposed, in this book, especially in the simulation part, to compare these two approaches based on their statistical AREs and computational performance under different innovation distributions mentioned in Chapter 1.

1. Conditional Maximum Likelihood

$$G_{Y_t} = \sum_{i_1=0}^{\min(Y_{t-1}, Y_t)} \binom{Y_{t-1}}{i_1} \rho_1^{i_1} (1-\rho_1)^{Y_{t-1}-i_1}$$

$$\times \sum_{i_2=0}^{\min(Y_{t-2}, Y_{t-i_1})} \binom{Y_{t-2}}{i_2} \rho_2^{i_2} (1-\rho_2)^{Y_{t-2}-i_2} \times \cdots$$

$$\times \sum_{i_p=0}^{\min(Y_{t-p}, Y_{t-(i_1+i_2+\cdots+i_p)})} \binom{Y_{t-p}}{i_p} \rho_p^{i_p} (1-\rho_p)^{Y_{t-p}-i_p}$$

$$\times f_{R_t}(Y_t - (i_1 + i_2 + \cdots + i_p)) \tag{2.13}$$

The conditional moments are easily derived from Equation 2.13 as follows:

$$E(Y_t | Y_{t-1}, Y_{t-2}, \ldots, Y_{t-p}) = E(R_t) + \sum_{k=1}^{p} \rho_k Y_{t-k};$$

$$Var(Y_t | Y_{t-1}, Y_{t-2}, \ldots, Y_{t-p}) = Var(R_t) + \sum_{l=k}^{p} \rho_k (1-\rho_k) Y_{t-l}.$$

From the joint PGF, assuming $\mathscr{F}_t = [Y_{t-1}, Y_{t-2}, \ldots, Y_{t-p}]$, we derive the cumulative conditional probability function as in Davies [1973] and Joe [2019], to obtain, $F_{Y_t | \mathscr{F}_t} = \frac{1}{2} - \frac{\pi}{2} \int_{-\pi}^{\pi} Re \left[\frac{G_{Y_t}(e^{i\omega}) \times e^{-i\omega Y_t}}{1 - e^{(-i\omega)}} \right] d\omega$ and, the conditional probability distribution is obtained as: $f_{y_t | \mathscr{F}_t} = F_{y_t | \mathscr{F}_t}(y_{t-1}) - F_{y_t | \mathscr{F}_t}(y_t)$ Let η be the vector of unknown parameters and the conditional likelihood is then constructed from: $L(\eta) = log \left[\prod_{t=2}^{T} f_{y_t | \mathscr{F}_t} \right]$ and $\hat{\eta}_{CML}$ is obtained by solving $\frac{\partial L(\eta)}{\partial \eta} = 0$, and from Bu et al. [2008], the same asymptotic properties hold such that, $\hat{\eta}_{CML} - \eta_0 \sim N \left(0, I_{CML}^{-1}(\hat{\eta}_{CML}) \right)$ where, η_0 is the true value.

2. The CLS estimating equation is expressed as:

$$Q(\eta) = \sum_{t=2}^{T} \left(Y_t - E(Y_t \mid Y_{t-1}, Y_{t-2}, \ldots, Y_{t-p}) \right)^2 \tag{2.14}$$

where, η is the set of unknown parameters and its true value is set to be η_0. In other words, $\hat{\eta}^{cls}$ is the solution of the score equation $\frac{\partial Q(\eta)}{\partial \eta} = 0$. For the 2-parameter distribution of R_t, Karlsen and Tjostheim [1988] suggested the 2-step CLS approach:

$$Q_T(\eta) = \sum_{t=2}^{T} \{ [Y_t - E(Y_t \mid Y_{t-1}, Y_{t-2}, \ldots, Y_{t-p})]^2$$

$$- Var(Y_t \mid Y_{t-1}, Y_{t-2}, \ldots, Y_{t-p}) \}^2 \tag{2.15}$$

Klimko and Nelson [1978] showed that for such CLS function, the $\hat{\eta}_{CLS}$ is strongly consistent and has an asymptotic distribution of the form:

$$\sqrt{T}(\eta^{cls} - \eta_0) \xrightarrow{d} N(\mathbf{0}, \mathbf{V}^{-1}\mathbf{W}\mathbf{V}^{-1}), \qquad (2.16)$$

where $\mathbf{W} = E\left(u_t(\eta)^2 \dfrac{\partial h_t(\eta)}{\partial \eta} \dfrac{\partial h_t(\eta)}{\partial \eta^\top}\right)_{\eta_0}$,

$\mathbf{V} = E\left(\dfrac{\partial h_t(\eta)}{\partial \eta} \dfrac{\partial h_t(\eta)}{\partial \eta^\top} - u_t(\eta)\dfrac{\partial^2 h_t(\eta)}{\partial \eta \partial \eta^\top}\right)_{\eta_0}$, $u_t(\eta) = Y_t - h_t(\eta)$, $h_t(\eta) = E(Y_t \mid Y_{t-1}, Y_{t-2}, \ldots, Y_{t-p})$.

In the context of the constrained BINAR (1), the conditional density for the constrained BINAR (1) process is expressed as:

$$f_1(y_1) = \binom{Y_{t-1,1}}{y_1} \rho_{11}^{y_1}(1 - \rho_{11})^{(y_{t-1,1}-y_1)}$$

$$f_2(y_2) = \binom{Y_{t-1,2}}{y_2} \rho_{22}^{y_2}(1 - \rho_{22})^{(y_{t-1,2}-y_2)}$$

and the joint conditional density is given by: $f(\tilde{y}_t \mid y_{t-1}^{\sim}, \tilde{\eta}) = \sum_k \sum_s f_1(y_{t,1} - k) \times f_2(y_{t,2} - s) \times P(R_{t,1} = k, R_{t,2} = s)$ and the conditional likelihood is then, $L(\eta \mid \tilde{y}_t) = \prod_{t=1}^T f(\tilde{y}_t \mid y_{t-1}^{\sim}, \tilde{\eta})$ where, $\tilde{y}_t = (y_{t,1}, y_{t,2})$ and the CLS [Nastic et al., 2016] is given as:

$$\Phi(\tilde{\eta}) = \sum_{t=1}^T (\tilde{y}_t - E(\tilde{y}_t \mid y_{t-1}^{\sim}))'(\tilde{y}_t - E(\tilde{y}_t \mid y_{t-1}^{\sim})) \qquad (2.17)$$

where,

$$E(\tilde{y}_t \mid y_{t-1}^{\sim}) = E\begin{bmatrix} Y_{t,1} \mid Y_{t-1,1} \\ Y_{t,2} \mid Y_{t-1,2} \end{bmatrix} = \begin{bmatrix} \rho_{11}Y_{t-1,1} + E(R_{t,1}) \\ \rho_{22}Y_{t-1,1} + E(R_{t,2}) \end{bmatrix}$$

under the Binomial thinning.

The probability distribution of $\{R_{t,1}, R_{t,2}\}$ may be specified with common time-varying covariates that influence $\{Y_{t,1}, Y_{t,2}\}$. In fact, in Mamode Khan et al. [2016], a comprehensive comparison study was conducted on the performance of the CML, CLS, and Quasi-Likelihood (QL) procedure, where the outcomes showed that CML and QL provide lesser biased estimates, with CML slightly better, but the QL approach showed lesser convergence failures.

For the unconstrained BINAR (1), again, the CML, CLS, and QL have been mostly used, with CML and QL yielding better efficient estimates and with QL having lesser computational failures. However, as of date, both the constrained and the unconstrained BINAR (1) models have been constructed for simple order 1 and with some innovative distributions that include only the Poisson, NB, and COM-Poisson marginals. There are yet some extended works that can be carried out to explore these processes for higher orders and with other generalized Poisson mixtures.

From Pedeli and Karlis [2013aa], the conditional likelihood is given by:

$$f_1(k) = \sum_{j_1=0}^{k} \binom{y_{t-1,1}}{j_1} \binom{y_{t-1,2}}{k-j_1}$$
$$\rho_{11}^{j_1}(1-\rho_{11})^{(y_{t-1,1}-j_1)} \times \rho_{12}^{k-j_1}(1-\rho_{12})^{(y_{t-1,2}-k+j_1)}$$

$$f_2(s) = \sum_{j_2=0}^{s} \binom{y_{t-1,2}}{j_2} \binom{y_{t-1,1}}{s-j_2}$$
$$\rho_{22}^{j_2}(1-\rho_{22})^{(y_{t-1,2}-j_2)} \times \rho_{21}^{s-j_2}(1-\rho_{21})^{(y_{t-1,1}-s+j_2)}$$

and, $f(\tilde{y}_t \mid y_{t-1}^{\sim}, \tilde{\eta}) = \sum_{k=0}^{g_1}\sum_{s=0}^{g_2} f_1(k)f_2(s)P(R_{t,1}=y_{t-k,1}, R_{t,2}=y_{t-s,2})$ and the CLS function is given by:

$$\Phi(\tilde{\eta}) - \sum_{t=1}^{T}(\tilde{y}_t - E(\tilde{y}_t \mid y_{t-1}^{\sim}))'(\tilde{y}_t - E(\tilde{y}_t \mid y_{t-1}^{\sim})) \tag{2.18}$$

while, $E(\tilde{y}_t \mid y_{t-1}^{\sim}) = E\begin{bmatrix} Y_{t,1} \mid Y_{t-1,1} \\ Y_{t,2} \mid Y_{t-1,2} \end{bmatrix} = \begin{bmatrix} \rho_{11}Y_{t-1,1}+\rho_{12}Y_{t-1,2}+E(R_{t,11}) \\ \rho_{21}Y_{t-1,1}+\rho_{22}Y_{t-1,2}+E(R_{t,22}) \end{bmatrix}$ under Binomial thinning.

The asymptotic properties of the CML approach for the BINAR (1) process have been discussed in Billingsley [1961a], Franke and Rao [1995], Pedeli and Karlis [2013aa], and that of CLS in Nastic et al. [2016].

2.5 Concluding Remarks

This chapter provides an extensive review on the INAR processes and their extensions to BINAR processes. The important highlights are how these INAR type processes can accommodate for the over-dispersion, periodicity, zero-inflation, non-stationarity in the series and the covariate specifications. The over-dispersion may be due to an amalgamation of all these features. We made mention of the different thinning mechanisms and the choice of the innovation distributions as a means to handle the over-dispersion. These reviews allow us to determine the potentials to explore new INAR processes with various class of Poisson-mixtures and their associated Zero-inflated or other type of inflated versions and the corresponding periodic INAR(p) (PINAR(p)) models. The extensions to the bivariate case become another important avenue of future exploration. Likewise, for the SARs-CoV-2 series, assuming $Y_{t,1}$ and $Y_{t,2}$ is the number of new COVID-19 infected or newly admitted cases and the corresponding number of death cases, a bivariate model of the following form may be considered, $Y_{t,1} = \rho_{11} * Y_{t-1,1} + R_{t,1}$ $Y_{t,2} = \rho_{21} * Y_{t-1,1} + \rho_{22} * Y_{t-1,2} + R_{t,2}$, $Cov(Y_{t,1},Y_{t,2}) = Cov(\rho_{11} * Y_{t-1,1}, \rho_{21} * Y_{t-1,1}) + Cov(\rho_{11} * Y_{t-1,1}, \rho_{21} * Y_{t-1,2}) + Cov(R_{t,1}, R_{t,2})$ and with no dependence on the innovation, $\{R_{t,1}, R_{t,2}\}$, the $Cov(Y_{t,1},Y_{t,2})$ can be obtained by the sum of the two components $Cov(\rho_{11} * Y_{t-1,1}, \rho_{21} * Y_{t-1,1})$ and $Cov(\rho_{11} * Y_{t-1,1}, \rho_{21} * Y_{t-1,2})$. Such a bivariate model can be extended to order p. Another highlight of this chapter is with regard to the estimation methods, where we lay emphasis on the CML and CLS approaches mainly. In the next chapter, we thus propose to explore some novel INAR processes with other Poisson mixture and zero-inflated distributed innovations and assess the behavior of the CML and CLS approaches under stationary and non-stationary conditions and under the different thinning.

3

Simulation Study

This chapter presents some numerical illustrations of the high-ordered INAR process with some Poisson mixed innovations and their zero-inflated associates. The Monte Carlo simulation experiment based on an INAR (4) process that is given as:

$$Y_t = \rho_1 * Y_{t-1} + \rho_2 * Y_{t-2} + \rho_3 * Y_{t-3} + \rho_4 * Y_{t-4} + R_t, \qquad (3.1)$$

where $*$ indicates the binomial thinning operator, with the $\rho = (0.1, 0.05, 0.2, 0.3)$ such that the sum of the entries in ρ is less than one and hence ensures stability of the INAR (4) process. We consider four scenarios of simulation, which are categorized into stationary and non-stationary innovation link functions as follows:

Stationary

1. $R_t \sim Poisson(\lambda_t = 0.9)$
2. $R_t \sim ZIPoisson(\lambda_t = 0.9, \pi_t = 0.9)$

Non-Stationary covariates

1. $R_t \sim Poisson(\lambda_t = \beta_0 + \beta_1 log(t))$
2. $R_t \sim ZIPoisson(\lambda_t = \beta_0 + \beta_1 log(t), \pi_t = \frac{\exp(\eta_0 + \eta_1 log(t))}{1 + \exp(\eta_0 + \eta_1 log(t))})$

with $\beta_0 = \eta_0 = 0.3$ and $\beta_1 = \eta_1 = 0.2$ and π_t controls the proportion of zeros in the simulated data. Further to the handling of over-dispersion in Chapter 2 in the INAR(p) process, via appropriate statistical tests, it has been verified and confirmed that the simulated series with the Poisson and Zero-inflated Poisson innovations, generated from *rzip* in library ZIM in R, yield an over-dispersed set of observations for the sizes ranging from $T = 100, 300, 500$. Thus, to further assess the consistency of the model estimates, it is proposed to fit the simulated INAR (4) with different Poisson mixed innovations summarized in Table 3.1. The CML and CLS estimation methods are used to estimate the parameters based on 150 replications.

Here, different conventional Poisson-mixture models will be assessed, but the most recent and less explored ones are the PWE and PGLD distributions. The reason to include them in our study is primarily based on the recent findings of Altun [2019] and Irshad et al. [2021a]. The former mentioned author brought forward the INAR (1) with PWE innovations and via its applicability to the Shots and Drugs data from the Police Car beat in Pittsburgh, concluded that the INAR (1) PWE performed better than other INAR (1) with Poisson Lindley and Generalized Poisson distributions. Similarly, in line with Abouammoh et al.'s [2015] works, Irshad et al. [2021a]

DOI: 10.1201/9781003677451-3

Innovation	Parameters	PGF
Poisson (P)	$\lambda_t > 0$	$G_{R_t}(s) = e^{\lambda_t(s-1)}$
Geometric	$\lambda_t^* \in (0,1), \lambda_t^* = \frac{\lambda_t}{\lambda_t+1}, \lambda_t > 0$	$G_{R_t}(s) = \frac{\lambda_t^* s}{1-(1-\lambda_t^*)s}$
Negative Binomial (NB)	$\lambda_t > 0, \nu > 0$	$G_{R_t}(s) = [1 + \nu\lambda_t(1-s)]^{-\nu^{-1}}$
Conway-Maxwell Poisson (CMP)	$\lambda_t > 0$ and $\nu > 0$	From Shmueli et al. [2005], Chatla and Shmueli [2018], Sellers et al. [2011], $G_{R_t}(s) = \frac{Z(\lambda_t s, \nu)}{Z(\lambda_t, \nu)}$, where, from Gaunt et al. [2016] $$Z(\lambda_t, \nu) = \frac{\exp(\nu\lambda_t^{1/\nu})}{\lambda_t^{(\nu-1)/2\nu}(2\pi)^{(\nu-1)/2}\sqrt{\nu}}\left(1 + c_1(\nu\lambda_t^{1/\nu})^{-1} + c_2(\nu\lambda_t^{1/\nu})^{-2} + O(\lambda_t^{-3/\lambda_t})\right)$$ $$c_1 = \frac{\nu^2-1}{24}$$ $$c_2 - c_1^2/2 = \frac{\nu^2-1}{48}$$
Poisson Lindley (PL)	$\lambda_t > 0$	From Ghitany and Al Mutairi [2009], Livio et al. [2018], Mamode Khan et al. [2019], $G_{R_t}(s) = \frac{\lambda_t^2}{\lambda_t+1}\frac{2+\lambda_t-s}{(\lambda_t+1-s)^2}$
Poisson-Tweedie (PT)	$a \geq 0$	From Bonat et al. [2017], El Shaarawia et al. [2010], $$G_{R_t}(s) = \begin{cases}\exp\left\{\frac{b_t}{a}\left((1-c_t)^a - (1-c_t s)^a\right)\right\}, & a \neq 0; \\ \left[\frac{(1-c_t)^{b_t}}{(1-c_t s)^{b_t}}\right], & a = 0.\end{cases}$$ The PGF can be re-parameterized with $b_t = \frac{\lambda_t(1-c_t)^{1-a}}{c_t}, c_t = \frac{D_t-1}{D_t-a}$.
Poisson-Inverse-Gaussian (PIG)	$a \geq 0$	Let $a = \frac{1}{2}$ in the above PGF of the PT models.
Cosine-Weighted-Geometric (WCG)	λ_t^* and $\nu, \lambda_t^* = \frac{\lambda_t}{1+\lambda_t}$ where $\lambda_t^* \in (0,1)$ and $\nu \in [0, \frac{\pi}{2})$	From Chesneau et al. [2020], Mamode Khan et al. [2021], $G_{R_t}(s) = \frac{c_{\lambda_t^*, \nu}}{2}\left[\frac{1-\lambda_t^{*2}s}{1-\lambda_t^* s} + \frac{1-\lambda_t^{*2}s\cos(2\nu)}{1-2\lambda_t^* s\cos(2\nu)+(\lambda_t^* s)^2}\right], s < -\ln(\lambda_t^*),$
Poisson-Weighted-Exponential (PWE)	$\lambda_t > 0, \nu > 0$	Referring to Altun [2019], $G_{R_t}(s) = \frac{\nu(\lambda_t+1)}{1-s+\nu(\lambda_t+1)}$
Poisson Generalized Lindley distribution (PGLD)	$\lambda_t > 0, \nu > 0$	Referring to Abouammoh et al.'s [2015], $G_{R_t}(s) = \frac{\lambda_t^\nu}{1+\lambda_t}\frac{2-s+\lambda_t}{(1-s+\lambda_t)^\nu}$

TABLE 3.1

Probability Generating Function for all distributions

came up with the two-parameter Poisson Generalized Lindley (TPPGLD) by mixing Poisson distribution with a new Generalized Lindley distribution. This novel discrete univariate distribution even proved reliable in modeling over-dispersed datasets. All these previous findings served as benchmark to include PWE and PGLD in our simulations especially when working with high orders, that is for high-order autoregressive processes, INAR(p). Below, we start by describing the probability generating function of each Poisson mixture models: Note the definition of λ_t and v are strictly different across these above models. Likewise, the λ_t specification in Poisson, NB, PT are distinct from that for CMP, PL, WCG, PWE and PGLD. v broadly indicates the variance or over-dispersion parameters or the trigonometric coefficient in the WCG.

3.1   The Different Zero-Inflated Poisson-Mixtures Models

Zero Inflated (ZI) models, introduced by Lambert [1992], are suitable for over-dispersed count data that exhibit excessive zeros. These data are commonly encountered in social sciences, likewise in the analysis of drug addicts [McCord and Ratcliffe, 2007], crimes [Ristic et al., 2009], adolescents' drinking patterns [Cranford et al., 2010], counselling session attendance [Chang and Saunders, 2002], or in the financial sectors such as in the modeling of insurance claims [Yip and Yau, 2005], and in health studies such as in dental caries [Javali and Pandit, 2010], in injection cessation in HIV patients [Mackesy-Amiti et al., 2011] and among many other applications areas mentioned in [Perumean-Chaney et al., 2012].

Basically the ZI models is a mixture of two distributions: Firstly, a probability distribution that degenerates at zero and mixed with a standard probability model such as the Poisson or NB model. The general form is given by: $P(R_t = r_t) = \pi g_1(r_t) + (1-\pi)g_2(r_t)$ where π, the mixing proportion lies in the interval between 0 and 1 and indicates the rate of zero inflation and $g_1(.)$ and $g_2(.)$ are the corresponding densities. By replacing $g_1(.)$ with a probability distribution that generates at zero and $g_2(.)$ by the Poisson distribution with parameter μ, we derive the ZI-P as: $P(R_t = 0) = \pi + (1-\pi)e^{-\mu_t}$, $P(R_t = r_t) = (1-\pi)\frac{\mu_t^{r_t}e^{-\mu_t}}{r_t!}$, $r_t = 1,2,3,\ldots$ and the corresponding probability generating function (PGF) is $G_{R_t}(s) = \pi + (1-\pi)\frac{e^{\mu_t s}}{e^{\mu_t}}$ $= \pi + (1-\pi)e^{\mu_t(s-1)}$. Similarly the ZI-Negative Binomial (ZI-NB) with parameter (μ_t, v^{-1}) is given by $P(R_t = 0) = \pi + (1-\pi)\left(\frac{v^{-1}}{\mu + v^{-1}}\right)^{v^{-1}}$, $P(R_t = r_t) = (1-\pi)\left\{\frac{\Gamma(r_t+v^{-1})}{r_t!\Gamma(v^{-1})}\left(\frac{v^{-1}}{\mu_t+v^{-1}}\right)^{v^{-1}}\left(\frac{\mu}{\mu+v^{-1}}\right)^{r_t}\right\}$, $r_t = 1,2,3,\ldots$ with PGF given as:

$G_{R_t}(s) = \pi + (1-\pi)[1+\nu\mu_t(1-s)]^{-\nu^{-1}}$; and, recently, [Sellers et al., 2016] proposed the ZI-COM-Poisson model (ZI-CMP) where, $P(R_t = 0) = \pi + (1-\pi)\frac{1}{Z(\lambda,\nu)}$, $P(R_t = r_t) = (1-\pi)\frac{1}{Z(\lambda,\nu)}\frac{\lambda^{r_t}}{(r_t!)^\nu}$, $r_t = 1,2,3,\ldots$ and its PGF is $G_{R_t}(s) = \pi + (1-\pi)\frac{Z(\lambda s,\nu)}{Z(\theta,\nu)}$, where the $Z(\lambda,\nu)$ from Gaunt et al. [2016] is already given in Table 3.1.

Next, the PGFs of the zero-inflated versions of the Geometric (ZI-Geometric), Poisson-Tweedie (ZI-PT), Poisson Lindley (ZI-PL), Cosine-Geometric models [Chesneau et al., 2020, Mamode Khan et al., 2021] (ZI-WCG), Poisson Weighted Exponential (ZI-PWE), and Poisson Generalized Lindley (ZI-PGLD), is simply given by $\pi + (1-\pi)G_{R_t}(s)$.

In the event, we have some explanatory variables, given by the vector $\tilde{x}_t$, which are known to influence the tth response variable $y = y_t$, then $\tilde{x}_t = [x_1,x_2,\ldots,x_p]^T$, and $\mu_t = \exp(x_t^T\tilde{\beta})$, for the tth term with $\beta = [\beta_1,\beta_2,\ldots,\beta_p]^T$. While for ZI-CMP, $\lambda_t = \exp(x_t^T\tilde{\beta})$ and $\kappa_t = \frac{\exp(x_t^T\tilde{\beta})}{1+\exp(x_t^T\tilde{\beta})}$.

Overall, for interested readers, the ZI data can easily be generated in R using $ifelse(rbinom(n,size=1,prob=\pi)>0,0,rdis(n,\lambda^*,\nu))$, where for the Poisson model, "rdis" is $rpois(n,\lambda=\mu)$, for NB model, "rdis" is $rnbinom(n, size=\frac{1}{\nu}, mean=\mu)$ and for COM-Poisson, "rdis" is $rcmp(n,\lambda,\nu)$. Similarly, for the PT model, refer to the *poistweedie* package in R or alternatively, the data can be obtained from ZIM [M.Yang et al., 2018], iZID [Wang et al., 2020], bZinb [Cho et al., 2019] and ZiC packages [Jochmann, 2017].

Under all the above-described models, we then conduct the simulation exercise to assess the performance of each proposed model. Results from Tables 3.2 to 3.15 refer.

Innovation	Method	T	ρ_1=0.1	ρ_2=0.05	ρ_3=0.2	ρ_4=0.3	λ_t
Poisson	CML	100	0.100 (0.005)	0.050 (0.003)	0.197 (0.001)	0.300 (0.004)	0.915 (0.001)
		300	0.101 (0.001)	0.049 (0.002)	0.201 (0.001)	0.306 (0.003)	0.935 (0.002)
		500	0.104 (0.000)	0.054 (0.001)	0.200 (0.000)	0.269 (0.003)	0.885 (0.001)
	CLS	100	0.095 (0.006)	0.048 (0.003)	0.196 (0.004)	0.311 (0.006)	0.883 (0.003)
		300	0.143 (0.005)	0.052 (0.002)	0.191 (0.003)	0.333 (0.004)	0.907 (0.002)
		500	0.161 (0.003)	0.051 (0.001)	0.213 (0.001)	0.314 (0.004)	0.897 (0.001)

TABLE 3.2

Simulated estimates and corresponding standard errors in () under Poisson innovation with binomial thinning, assuming stationary setting

Innovation	Method	T	ρ_1=0.1	ρ_2=0.05	ρ_3=0.2	ρ_4=0.3	β_0=0.3	β_1=0.2
Poisson	CML	100	0.103 (0.013)	0.054 (0.012)	0.178 (0.009)	0.302 (0.009)	0.282 (0.003)	0.236 (0.005)
		300	0.110 (0.011)	0.047 (0.009)	0.210 (0.006)	0.303 (0.006)	0.271 (0.003)	0.242 (0.006)
		500	0.095 (0.008)	0.068 (0.007)	0.215 (0.006)	0.319 (0.005)	0.337 (0.002)	0.184 (0.003)
	CLS	100	0.107 (0.016)	0.049 (0.013)	0.164 (0.011)	0.331 (0.011)	0.348 (0.007)	0.236 (0.008)
		300	0.126 (0.017)	0.043 (0.014)	0.185 (0.010)	0.324 (0.008)	0.306 (0.006)	0.196 (0.005)
		500	0.135 (0.009)	0.044 (0.006)	0.243 (0.008)	0.264 (0.007)	0.296 (0.004)	0.217 (0.001)

TABLE 3.3
Simulated estimates and corresponding standard errors in () under Poisson innovation
with binomial thinning, assuming non-stationary setting

Innovation	Method	T	ρ_1=0.1	ρ_2=0.05	ρ_3=0.2	ρ_4=0.3	λ_t	π_t
ZI-P	CML	100	0.117 (0.003)	0.049 (0.003)	0.279 (0.002)	0.304 (0.003)	0.867 (0.004)	0.871 (0.001)
		300	0.142 (0.002)	0.055 (0.002)	0.195 (0.001)	0.300 (0.001)	0.932 (0.002)	0.896 (0.000)
		500	0.158 (0.001)	0.062 (0.003)	0.222 (0.000)	0.275 (0.000)	1.002 (0.001)	0.888 (0.000)
	CLS	100	0.164 (0.006)	0.051 (0.003)	0.245 (0.004)	0.286 (0.003)	0.999 (0.006)	0.832 (0.003)
		300	0.112 (0.004)	0.045 (0.002)	0.200 (0.003)	0.306 (0.002)	0.861 (0.005)	0.868 (0.001)
		500	0.136 (0.003)	0.063 (0.001)	0.237 (0.001)	0.332 (0.001)	1.036 (0.003)	0.866 (0.001)

TABLE 3.4
Simulated estimates and corresponding standard errors in () under Zero-Inflated Poisson (ZI-P) innovation with binomial thinning, assuming stationary setting

3.2 Numerical Solutions for Different Innovations under Binomial Thinning

The results from Tables 3.2 to 3.9 show that the simulated mean estimates of the
model parameters on the 150 replications, in particular for the ρ and β_0, β_1, η_0, η_1
are consistent with their corresponding population parameters. In fact, by comparing
the biases, we note that the INAR (4) with NB, PWE, PGLD and Geometric models
and with their zero-inflated associates under the binomial thinning operator provides
estimates with lower standard errors than the other competitive INAR (4) under both

Innovation	Method	T	$\rho_1=0.1$	$\rho_2=0.05$	$\rho_3=0.2$	$\rho_4=0.3$	$\beta_0=0.3$	$\beta_1=0.2$	$\eta_0=0.3$	$\eta_1=0.2$
ZI-P	CML	100	0.125 (0.003)	0.043 (0.004)	0.228 (0.003)	0.304 (0.007)	0.286 (0.003)	0.201 (0.005)	0.300 (0.008)	0.222 (0.000)
		300	0.100 (0.002)	0.054 (0.003)	0.196 (0.002)	0.199 (0.005)	0.322 (0.002)	0.248 (0.003)	0.315 (0.007)	0.201 (0.000)
		500	0.142 (0.001)	0.052 (0.001)	0.243 (0.001)	0.257 (0.004)	0.316 (0.001)	0.189 (0.001)	0.297 (0.005)	0.238 (0.000)
	CLS	100	0.117 (0.007)	0.049 (0.008)	0.196 (0.005)	0.265 (0.009)	0.271 (0.008)	0.211 (0.007)	0.277 (0.0011)	0.174 (0.003)
		300	0.106 (0.005)	0.046 (0.003)	0.224 (0.004)	0.338 (0.006)	0.325 (0.004)	0.227 (0.004)	0.335 (0.007)	0.245 (0.000)
		500	0.098 (0.004)	0.051 (0.002)	0.197 (0.002)	0.344 (0.003)	0.277 (0.003)	0.197 (0.002)	0.331 (0.005)	0.230 (0.000)

TABLE 3.5
Simulated estimates and corresponding standard errors in () under Zero-Inflated Poisson (ZI-P) innovation with binomial thinning, assuming non-stationary setting

stationary and non-stationary covariates. Most importantly, we observed smooth execution of simulations of the mentioned best suited models under binomial thinning operator and under the stationary and non-stationary settings, with no computational failures. Next, as the number of time points increases, the standard errors of the CML and CLS estimates decrease, which is as expected. As for the estimates of v, we cannot comment on their consistencies, since the simulated data are from an INAR (4) with Poisson innovations but the v estimates from the different models justify the over-dispersion in the data. In terms of the non-convergence simulations, in particular with the covariates and the zero-inflated versions, we notice that the CML for the INAR (4) with CMP, WCG, PT, and PIG innovations for $T = 300$ and $T = 500$ are very time consuming and at $T = 300$, around 4% and 6% simulation fail, with the *optim* routine for both distributions. However, under the CLS approach, we do not obtain any convergence failures, but CML yield slightly lower standard errors.

3.3 Numerical Solutions for Different Innovations under Generalized Binomial Thinning

Simulation procedures were also run for the different Poisson-mixture models, under the Generalized Binomial thinning. Based on 150 replications, the results are displayed in Tables 3.10-3.17.

We notice from the Tables 3.10 to 3.17 that both CML and CLS provide consistent simulated mean estimates and with CML providing slightly better standard errors than CLS and hence CML has better AREs. Still, under the Generalized Binomial thinning procedure, the best three models—INAR (4) with NB, PWE and

Innovation	Method	T	$\rho_1 = 0.1$	ρ_2=0.05	ρ_3=0.2	ρ_4=0.3	λ_t	v
Geometric	CML	100	0.093	0.049	0.200	0.332	0.883	
			(0.002)	(0.002)	(0.003)	(0.001)	(0.000)	
		300	0.162	0.045	0.197	0.326	0.913	
			(0.002)	(0.001)	(0.002)	(0.000)	(0.000)	
		500	0.101	0.056	0.223	0.300	0.947	
			(0.001)	(0.000)	(0.001)	(0.000)	(0.000)	
	CLS	100	0.095	0.053	0.186	0.289	0.923	
			(0.003)	(0.004)	(0.003)	(0.002)	(0.001)	
		300	0.112	0.046	0.177	0.322	0.896	
			(0.002)	(0.003)	(0.003)	(0.001)	(0.000)	
		500	0.103	0.041	0.201	0.308	0.911	
			(0.002)	(0.001)	(0.002)	(0.000)	(0.000)	
NB	CML	100	0.102	0.062	0.211	0.321	0.883	0.884
			(0.003)	(0.003)	(0.002)	(0.004)	(0.000)	(0.000)
		300	0.134	0.050	0.235	0.259	0.913	0.862
			(0.001)	(0.002)	(0.001)	(0.003)	(0.000)	(0.000)
		500	0.114	0.047	0.215	0.277	0.947	0.996
			(0.000)	(0.001)	(0.001)	(0.001)	(0.000)	(0.000)
	CLS	100	0.086	0.053	0.186	0.266	0.906	0.938
			(0.003)	(0.005)	(0.003)	(0.004)	(0.002)	(0.002)
		300	0.096	0.047	0.202	0.324	0.876	0.873
			(0.002)	(0.003)	(0.002)	(0.003)	(0.002)	(0.001)
		500	0.122	0.049	0.199	0.268	0.913	0.912
			(0.001)	(0.002)	(0.002)	(0.002)	(0.000)	(0.000)
CMP	CML	100	0.128	0.056	0.186	0.323	1.056	0.812
			(0.021)	(0.017)	(0.008)	(0.011)	(0.009)	(0.009)
		300	0.113	0.047	0.222	0.286	1.129	0.726
			(0.018)	(0.013)	(0.005)	(0.008)	(0.008)	(0.005)
		500	0.095	0.051	0.197	0.310	1.654	0.811
			(0.015)	(0.011)	(0.004)	(0.006)	(0.006)	(0.003)
	CLS	100	0.125	0.049	0.165	0.336	1.326	0.859
			(0.025)	(0.023)	(0.006)	(0.015)	(0.011)	(0.010)
		300	0.134	0.053	0.233	0.278	1.478	0.722
			(0.019)	(0.021)	(0.005)	(0.013)	(0.008)	(0.008)
		500	0.098	0.055	0.215	0.336	1.885	0.803
			(0.016)	(0.018)	(0.004)	(0.009)	(0.005)	(0.005)
PL	CML	100	0.100	0.070	0.233	0.274	1.153	
			(0.006)	(0.019)	(0.004)	(0.006)	(0.004)	
		300	0.098	0.043	0.188	0.286	1.085	
			(0.005)	(0.011)	(0.003)	(0.005)	(0.003)	
		500	0.122	0.057	0.169	0.368	1.526	
			(0.003)	(0.009)	(0.002)	(0.003)	(0.002)	
	CLS	100	0.132	0.053	0.241	0.326	1.027	
			(0.009)	(0.023)	(0.008)	(0.010)	(0.006)	
		300	0.090	0.045	0.168	0.297	1.226	
			(0.007)	(0.018)	(0.005)	(0.008)	(0.005)	
		500	0.119	0.056	0.211	0.305	1.583	
			(0.005)	(0.012)	(0.003)	(0.005)	(0.003)	

Innovation	Method	T	$\rho_1 = 0.1$	$\rho_2=0.05$	$\rho_3=0.2$	$\rho_4=0.3$	λ_t	v
PT	CML	100	0.145	0.060	0.199	0.287	0.896	1.023
			(0.017)	(0.009)	(0.004)	(0.008)	(0.011)	(0.006)
		300	0.098	0.051	0.235	0.264	0.865	1.121
			(0.015)	(0.003)	(0.002)	(0.006)	(0.009)	(0.004)
		500	0.100	0.049	0.228	0.325	0.799	1.687
			(0.011)	(0.002)	(0.001)	(0.004)	(0.007)	(0.002)
	CLS	100	0.115	0.053	0.196	0.338	0.943	1.127
			(0.016)	(0.011)	(0.006)	(0.005)	(0.009)	(0.009)
		300	0.086	0.039	0.222	0.259	0.865	1.223
			(0.009)	(0.008)	(0.005)	(0.003)	(0.007)	(0.006)
		500	0.153	0.048	0.185	0.307	0.916	1.428
			(0.008)	(0.005)	(0.003)	(0.001)	(0.004)	(0.005)
PIG	CML	100	0.097	0.062	0.187	0.387	0.884	
			(0.027)	(0.018)	(0.000)	(0.006)	(0.005)	
		300	0.149	0.057	0.196	0.402	0.902	
			(0.021)	(0.009)	(0.002)	(0.005)	(0.003)	
		500	0.142	0.068	0.154	0.321	0.861	
			(0.018)	(0.005)	(0.001)	(0.004)	(0.002)	
	CLS	100	0.120	0.055	0.188	0.325	0.865	
			(0.029)	(0.015)	(0.005)	(0.008)	(0.005)	
		300	0.086	0.048	0.229	0.314	0.928	
			(0.020)	(0.011)	(0.002)	(0.005)	(0.002)	
		500	0.136	0.053	0.176	0.266	0.906	
			(0.018)	(0.009)	(0.000)	(0.003)	(0.001)	
WCG	CML	100	0.102	0.045	0.266	0.317	1.008	1.325
			(0.003)	(0.002)	(0.002)	(0.002)	(0.004)	(0.003)
		300	0.098	0.057	0.189	0.348	1.745	1.258
			(0.002)	(0.001)	(0.001)	(0.001)	(0.003)	(0.002)
		500	0.125	0.055	0.214	0.301	1.459	1.467
			(0.000)	(0.001)	(0.000)	(0.000)	(0.001)	(0.001)
	CLS	100	0.095	0.063	0.223	0.326	0.916	1.118
			(0.005)	(0.003)	(0.004)	(0.004)	(0.005)	(0.002)
		300	0.130	0.057	0.196	0.286	0.864	1.325
			(0.004)	(0.002)	(0.003)	(0.002)	(0.004)	(0.001)
		500	0.142	0.046	0.211	0.305	0.932	1.765
			(0.001)	(0.001)	(0.002)	(0.000)	(0.002)	(0.000)
PWE	CML	100	0.121	0.051	0.235	0.342	1.664	0.875
			(0.000)	(0.001)	(0.000)	(0.000)	(0.001)	(0.000)
		300	0.136	0.049	0.184	0.284	1.285	0.904
			(0.000)	(0.000)	(0.000)	(0.000)	(0.000)	(0.000)
		500	0.094	0.050	0.223	0.304	1.078	0.896
			(0.000)	(0.000)	(0.000)	(0.000)	(0.000)	(0.000)
	CLS	100	0.098	0.053	0.223	0.329	1.232	0.926
			(0.001)	(0.000)	(0.001)	(0.002)	(0.001)	(0.001)
		300	0.146	0.048	0.196	0.299	1.442	0.969
			(0.000)	(0.000)	(0.000)	(0.001)	(0.000)	(0.000)
		500	0.132	0.050	0.214	0.307	1.628	0.887
			(0.000)	(0.000)	(0.000)	(0.000)	(0.000)	(0.000)
PGLD	CML	100	0.125	0.045	0.213	0.342	0.905	1.002
			(0.000)	(0.000)	(0.000)	(0.000)	(0.000)	(0.000)
		300	0.096	0.052	0.189	0.289	0.889	0.986
			(0.000)	(0.000)	(0.000)	(0.000)	(0.000)	(0.000)
		500	0.112	0.053	0.206	0.336	0.977	0.889
			(0.000)	(0.000)	(0.000)	(0.000)	(0.000)	(0.000)
	CLS	100	0.142	0.048	0.196	0.309	0.806	0.955
			(0.001)	(0.001)	(0.000)	(0.000)	(0.002)	(0.000)
		300	0.107	0.051	0.222	0.297	0.816	0.997
			(0.000)	(0.001)	(0.000)	(0.000)	(0.000)	(0.000)
		500	0.099	0.055	0.213	0.306	0.931	0.964
			(0.000)	(0.000)	(0.000)	(0.000)	(0.000)	(0.000)

TABLE 3.6

Simulated estimates and corresponding standard errors in () under different Poisson-mixture innovations with binomial thinning, assuming stationarity

Innovation	Method	T	$\rho_1=0.1$	$\rho_2=0.05$	$\rho_3=0.2$	$\rho_4=0.3$	$\beta_0=0.3$	$\beta_1=0.2$	ν
Geometric	CML	100	0.112	0.049	0.200	0.332	0.338	0.189	
			(0.002)	(0.002)	(0.003)	(0.001)	(0.011)	(0.010)	
		300	0.162	0.045	0.197	0.326	0.277	0.214	
			(0.002)	(0.001)	(0.002)	(0.000)	(0.009)	(0.006)	
		500	0.101	0.056	0.223	0.300	0.316	0.236	
			(0.001)	(0.000)	(0.001)	(0.000)	(0.005)	(0.002)	
	CLS	100	0.086	0.053	0.186	0.326	0.302	0.236	
			(0.003)	(0.004)	(0.002)	(0.002)	(0.009)	(0.013)	
		300	0.124	0.048	0.212	0.286	0.296	0.199	
			(0.002)	(0.003)	(0.001)	(0.001)	(0.006)	(0.009)	
		500	0.112	0.052	0.198	0.332	0.311	0.216	
			(0.001)	(0.002)	(0.000)	(0.000)	(0.005)	(0.006)	
NB	CML	100	0.121	0.052	0.218	0.328	0.318	0.199	0.699
			(0.018)	(0.011)	(0.016)	(0.008)	(0.010)	(0.008)	(0.005)
		300	0.088	0.049	0.226	0.289	0.326	0.214	0.841
			(0.012)	(0.008)	(0.012)	(0.006)	(0.009)	(0.005)	(0.004)
		500	0.132	0.047	0.189	0.304	0.289	0.226	0.776
			(0.009)	(0.005)	(0.009)	(0.003)	(0.007)	(0.003)	(0.001)
	CLS	100	0.099	0.055	0.231	0.336	0.306	0.178	0.829
			(0.020)	(0.016)	(0.023)	(0.009)	(0.016)	(0.009)	(0.009)
		300	0.104	0.056	0.196	0.265	0.333	0.193	0.965
			(0.019)	(0.015)	(0.018)	(0.005)	(0.012)	(0.007)	(0.007)
		500	0.086	0.048	0.211	0.229	0.287	0.223	0.815
			(0.017)	(0.012)	(0.016)	(0.003)	(0.010)	(0.004)	(0.002)
CMP	CML	100	0.099	0.048	0.222	0.294	0.348	0.234	0.786
			(0.021)	(0.012)	(0.009)	(0.010)	(0.016)	(0.021)	(0.004)
		300	0.158	0.056	0.218	0.285	0.341	0.248	0.871
			(0.015)	(0.010)	(0.007)	(0.006)	(0.012)	(0.019)	(0.002)
		500	0.086	0.042	0.199	0.267	0.351	0.243	0.899
			(0.012)	(0.011)	(0.005)	(0.004)	(0.010)	(0.016)	(0.001)
	CLS	100	0.105	0.054	0.225	0.329	0.342	0.223	0.863
			(0.030)	(0.010)	(0.011)	(0.012)	(0.020)	(0.019)	(0.005)
		300	0.096	0.046	0.230	0.332	0.264	0.188	0.789
			(0.025)	(0.008)	(0.008)	(0.009)	(0.015)	(0.015)	(0.003)
		500	0.127	0.049	0.196	0.256	0.300	0.242	0.886
			(0.018)	(0.007)	(0.005)	(0.005)	(0.009)	(0.013)	(0.816)
PL	CML	100	0.099	0.052	0.301	0.288	0.311	0.254	
			(0.035)	(0.022)	(0.018)	(0.012)	(0.027)	(0.011)	
		300	0.178	0.036	0.266	0.325	0.337	0.189	
			(0.031)	(0.018)	(0.021)	(0.009)	(0.025)	(0.009)	
		500	0.111	0.047	0.256	0.411	0.284	0.223	
			(0.022)	(0.015)	(0.017)	(0.005)	(0.019)	(0.004)	
	CLS	100	0.105	0.056	0.231	0.337	0.258	0.188	
			(0.036)	(0.025)	(0.015)	(0.009)	(0.030)	(0.016)	
		300	0.099	0.049	0.186	0.259	0.304	0.227	
			(0.029)	(0.019)	(0.013)	(0.005)	(0.028)	(0.014)	
		500	0.127	0.052	0.246	0.297	0.267	0.206	
			(0.025)	(0.015)	(0.009)	(0.003)	(0.019)	(0.011)	

Innovation	Method	T	$\rho_1=0.1$	$\rho_2=0.05$	$\rho_3=0.2$	$\rho_4=0.3$	$\beta_0=0.3$	$\beta_1=0.2$	v
PT	CML	100	0.134	0.053	0.188	0.326	0.286	0.211	1.078
			(0.030)	(0.021)	(0.009)	(0.015)	(0.012)	(0.008)	(0.004)
		300	0.099	0.049	0.228	0.278	0.304	0.167	1.799
			(0.026)	(0.016)	(0.006)	(0.012)	(0.009)	(0.006)	(0.003)
		500	0.105	0.052	0.223	0.338	0.333	0.188	1.421
			(0.022)	(0.014)	(0.004)	(0.009)	(0.008)	(0.004)	(0.001)
	CLS	100	0.081	0.044	0.178	0.313	0.328	0.231	1.223
			(0.031)	(0.018)	(0.008)	(0.016)	(0.017)	(0.009)	(0.004)
		300	0.121	0.060	0.201	0.294	0.268	0.246	1.529
			(0.029)	(0.016)	(0.005)	(0.009)	(0.016)	(0.007)	(0.003)
		500	0.098	0.058	0.196	0.277	0.344	0.186	1.845
			(0.026)	(0.014)	(0.004)	(0.005)	(0.012)	(0.005)	(0.002)
PIG	CML	100	0.121	0.058	0.187	0.335	0.266	0.199	
			(0.018)	(0.011)	(0.009)	(0.003)	(0.014)	(0.005)	
		300	0.148	0.057	0.228	0.287	0.364	0.175	
			(0.009)	(0.011)	(0.010)	(0.005)	(0.011)	(0.003)	
		500	0.115	0.052	0.185	0.316	0.356	0.188	
			(0.005)	(0.006)	(0.008)	(0.003)	(0.008)	(0.001)	
	CLS	100	0.132	0.053	0.198	0.326	0.259	0.223	
			(0.020)	(0.015)	(0.008)	(0.004)	(0.011)	(0.007)	
		300	0.096	0.044	0.223	0.267	0.288	0.166	
			(0.018)	(0.012)	(0.006)	(0.003)	(0.008)	(0.006)	
		500	0.117	0.062	0.166	0.305	0.311	0.208	
			(0.012)	(0.008)	(0.005)	(0.001)	(0.006)	(0.004)	
WCG	CML	100	0.087	0.059	0.223	0.342	0.331	0.236	1.223
			(0.008)	(0.008)	(0.013)	(0.011)	(0.008)	(0.011)	(0.003)
		300	0.095	0.040	0.197	0.336	0.287	0.228	1.412
			(0.007)	(0.006)	(0.012)	(0.009)	(0.005)	(0.009)	(0.002)
		500	0.126	0.057	0.216	0.269	0.336	0.247	1.501
			(0.006)	(0.004)	(0.009)	(0.006)	(0.003)	(0.006)	(0.001)
	CLS	100	0.126	0.056	0.235	0.321	0.264	0.264	1.232
			(0.008)	(0.007)	(0.018)	(0.009)	(0.010)	(0.012)	(0.005)
		300	0.096	0.048	0.186	0.318	0.316	0.189	1.448
			(0.005)	(0.005)	(0.016)	(0.006)	(0.009)	(0.009)	(0.004)
		500	0.122	0.051	0.221	0.289	0.333	0.217	1.634
			(0.003)	(0.004)	(0.012)	(0.005)	(0.007)	(0.005)	(0.003)
PWE	CML	100	0.119	0.046	0.275	0.307	0.314	0.232	0.852
			(0.001)	(0.000)	(0.000)	(0.001)	(0.000)	(0.001)	(0.002)
		300	0.090	0.043	0.203	0.336	0.365	0.196	0.896
			(0.000)	(0.000)	(0.000)	(0.000)	(0.000)	(0.000)	(0.001)
		500	0.123	0.056	0.287	0.268	0.346	0.243	0.901
			(0.000)	(0.000)	(0.000)	(0.000)	(0.000)	(0.000)	(0.000)
	CLS	100	0.134	0.051	0.321	0.329	0.346	0.199	0.856
			(0.002)	(0.001)	(0.000)	(0.002)	(0.001)	(0.000)	(0.002)
		300	0.095	0.046	0.296	0.278	0.289	0.246	0.857
			(0.001)	(0.000)	(0.000)	(0.001)	(0.000)	(0.000)	(0.001)
		500	0.123	0.058	0.314	0.322	0.333	0.231	0.899
			(0.000)	(0.000)	(0.000)	(0.000)	(0.000)	(0.000)	(0.000)
PGLD	CML	100	0.104	0.053	0.209	0.326	0.306	0.234	0.908
			(0.000)	(0.000)	(0.000)	(0.000)	(0.000)	(0.000)	(0.000)
		300	0.136	0.048	0.196	0.268	0.324	0.196	0.889
			(0.000)	(0.000)	(0.000)	(0.000)	(0.000)	(0.000)	(0.000)
		500	0.116	0.059	0.212	0.307	0.320	0.208	0.943
			(0.000)	(0.000)	(0.000)	(0.000)	(0.000)	(0.000)	(0.000)
	CLS	100	0.089	0.052	0.246	0.266	0.295	0.198	0.898
			(0.000)	(0.000)	(0.000)	(0.000)	(0.000)	(0.000)	(0.000)
		300	0.137	0.050	0.199	0.341	0.311	0.201	0.926
			(0.000)	(0.000)	(0.000)	(0.000)	(0.000)	(0.000)	(0.000)
		500	0.128	0.049	0.228	0.299	0.325	0.219	1.021
			(0.000)	(0.000)	(0.000)	(0.000)	(0.000)	(0.000)	(0.000)

TABLE 3.7
Simulated estimates and corresponding standard errors in () under different Poisson-mixture innovations with binomial thinning, assuming non-stationarity

Innovation	Method	T	ρ_1=0.1	ρ_2=0.05	ρ_3=0.2	ρ_4=0.3	λ_t	π_t	ν
ZI-Geometric	CML	100	0.101	0.046	0.202	0.289	0.786	0.889	
			(0.008)	(0.006)	(0.004)	(0.005)	(0.004)	(0.004)	
		300	0.125	0.052	0.196	0.344	0.812	0.856	
			(0.006)	(0.005)	(0.003)	(0.003)	(0.003)	(0.003)	
		500	0.102	0.049	0.235	0.342	0.896	0.847	
			(0.005)	(0.003)	(0.002)	(0.002)	(0.001)	(0.002)	
	CLS	100	0.086	0.053	0.232	0.336	0.925	0.927	
			(0.010)	(0.007)	(0.003)	(0.006)	(0.008)	(0.006)	
		300	0.123	0.046	0.196	0.298	0.889	0.864	
			(0.008)	(0.005)	(0.002)	(0.005)	(0.005)	(0.004)	
		500	0.098	0.055	0.158	0.302	0.912	0.900	
			(0.007)	(0.002)	(0.001)	(0.004)	(0.004)	(0.003)	
ZI-NB	CML	100	0.095	0.051	0.213	0.289	0.873	0.896	0.916
			(0.010)	(0.012)	(0.013)	(0.009)	(0.008)	(0.005)	(0.003)
		300	0.132	0.046	0.186	0.302	0.901	0.854	0.857
			(0.008)	(0.008)	(0.014)	(0.008)	(0.006)	(0.003)	(0.002)
		500	0.118	0.053	0.176	0.277	0.865	0.881	0.922
			(0.006)	(0.006)	(0.012)	(0.006)	(0.005)	(0.002)	(0.001)
	CLS	100	0.118	0.052	0.218	0.332	0.941	0.911	0.942
			(0.012)	(0.015)	(0.011)	(0.009)	(0.011)	(0.008)	(0.005)
		300	0.086	0.049	0.167	0.296	0.867	0.867	0.886
			(0.009)	(0.009)	(0.009)	(0.007)	(0.008)	(0.007)	(0.003)
		500	0.107	0.058	0.233	0.311	0.914	0.927	0.859
			(0.007)	(0.005)	(0.005)	(0.005)	(0.005)	(0.005)	(0.006)
ZI-CMP	CML	100	0.092	0.054	0.227	0.301	1.002	0.899	0.756
			(0.012)	(0.016)	(0.011)	(0.011)	(0.009)	(0.011)	(0.005)
		300	0.096	0.045	0.205	0.296	1.126	0.864	0.877
			(0.010)	(0.014)	(0.009)	(0.009)	(0.006)	(0.008)	(0.004)
		500	0.184	0.053	0.198	0.264	1.247	0.891	0.896
			(0.005)	(0.012)	(0.005)	(0.006)	(0.004)	(0.006)	(0.004)
	CLS	100	0.114	0.044	0.231	0.341	1.314	0.926	0.849
			(0.016)	(0.014)	(0.013)	(0.008)	(0.012)	(0.013)	(0.003)
		300	0.099	0.056	0.168	0.296	1.465	0.865	0.923
			(0.014)	(0.012)	(0.015)	(0.005)	(0.009)	(0.011)	(0.002)
		500	0.121	0.049	0.229	0.316	1.886	0.907	0.877
			(0.011)	(0.011)	(0.009)	(0.002)	(0.008)	(0.008)	(0.001)
ZI-PL	CML	100	0.129	0.061	0.223	0.311	1.325	0.886	
			(0.025)	(0.016)	(0.023)	(0.015)	(0.009)	(0.006)	
		300	0.132	0.055	0.274	0.326	1.246	0.877	
			(0.021)	(0.014)	(0.019)	(0.013)	(0.006)	(0.005)	
		500	0.119	0.059	0.176	0.333	1.122	0.884	
			(0.019)	(0.012)	(0.016)	(0.009)	(0.005)	(0.004)	
	CLS	100	0.111	0.048	0.222	0.279	1.642	0.921	
			(0.030)	(0.015)	(0.025)	(0.017)	(0.008)	(0.009)	
		300	0.100	0.053	0.237	0.321	1.441	0.886	
			(0.027)	(0.013)	(0.019)	(0.015)	(0.005)	(0.005)	
		500	0.865	0.059	0.248	0.288	1.023	0.912	
			(0.023)	(0.009)	(0.015)	(0.011)	(0.003)	(0.004)	

Innovation	Method	T	$\rho_1=0.1$	$\rho_2=0.05$	$\rho_3=0.2$	$\rho_4=0.3$	λ_t	π_t	ν
ZI-PT	CML	100	0.147 (0.024)	0.039 (0.019)	0.222 (0.016)	0.282 (0.017)	0.902 (0.021)	0.795 (0.018)	1.269 (0.003)
		300	0.111 (0.021)	0.046 (0.018)	0.274 (0.014)	0.286 (0.016)	0.877 (0.018)	0.862 (0.015)	1.505 (0.002)
		500	0.098 (0.019)	0.052 (0.017)	0.210 (0.010)	0.315 (0.014)	0.854 (0.015)	0.901 (0.013)	1.801 (0.001)
	CLS	100	0.156 (0.031)	0.056 (0.016)	0.231 (0.018)	0.321 (0.019)	0.896 (0.025)	0.867 (0.015)	1.235 (0.006)
		300	0.132 (0.025)	0.047 (0.012)	0.246 (0.015)	0.289 (0.015)	0.902 (0.019)	0.855 (0.013)	1.647 (0.004)
		500	0.102 (0.099)	0.051 (0.009)	0.186 (0.012)	0.323 (0.013)	0.889 (0.015)	0.911 (0.009)	1.526 (0.003)
ZI-PIG	CML	100	0.146 (0.018)	0.042 (0.009)	0.196 (0.016)	0.301 (0.018)	0.905 (0.021)	0.8896 (0.022)	
		300	0.096 (0.016)	0.044 (0.010)	0.217 (0.014)	0.265 (0.016)	0.889 (0.019)	0.843 (0.019)	
		500	0.105 (0.008)	0.058 (0.004)	0.196 (0.011)	0.293 (0.011)	0.875 (0.018)	0.912 (0.016)	
	CLS	100	0.096 (0.022)	0.050 (0.012)	0.203 (0.014)	0.346 (0.021)	0.921 (0.025)	0.977 (0.026)	
		300	0.128 (0.016)	0.049 (0.011)	0.179 (0.013)	0.290 (0.018)	0.859 (0.018)	0.896 (0.013)	
		500	0.099 (0.012)	0.052 (0.009)	0.222 (0.011)	0.309 (0.015)	0.913 (0.015)	0.903 (0.009)	
ZI-WCG	CML	100	0.096 (0.005)	0.056 (0.004)	0.186 (0.006)	0.268 (0.005)	1.025 (0.007)	0.866 (0.003)	1.002 (0.003)
		300	0.108 (0.003)	0.041 (0.003)	0.214 (0.005)	0.278 (0.003)	1.116 (0.005)	0.809 (0.002)	1.245 (0.002)
		500	0.124 (0.002)	0.048 (0.001)	0.261 (0.004)	0.324 (0.001)	1.245 (0.004)	0.902 (0.001)	1.406 (0.001)
	CLS	100	0.128 (0.009)	0.056 (0.006)	0.232 (0.005)	0.311 (0.008)	1.416 (0.009)	0.923 (0.005)	0.789 (0.006)
		300	0.089 (0.007)	0.048 (0.005)	0.196 (0.003)	0.267 (0.006)	1.235 (0.007)	0.867 (0.004)	1.035 (0.005)
		500	0.106 (0.005)	0.051 (0.003)	0.187 (0.002)	0.301 (0.005)	1.006 (0.005)	0.921 (0.003)	0.965 (0.004)
ZI-PWE	CML	100	0.103 (0.002)	0.050 (0.001)	0.227 (0.002)	0.352 (0.001)	1.075 (0.002)	0.845 (0.001)	0.854 (0.002)
		300	0.114 (0.001)	0.057 (0.000)	0.206 (0.001)	0.269 (0.000)	1.186 (0.001)	0.789 (0.000)	0.876 (0.001)
		500	0.132 (0.000)	0.054 (0.000)	0.200 (0.000)	0.339 (0.000)	1.201 (0.000)	0.883 (0.000)	0.897 (0.000)
	CLS	100	0.123 (0.003)	0.048 (0.002)	0.249 (0.001)	0.322 (0.003)	1.112 (0.003)	0.877 (0.002)	0.916 (0.001)
		300	0.098 (0.001)	0.046 (0.001)	0.199 (0.000)	0.346 (0.002)	1.446 (0.000)	0.912 (0.001)	0.899 (0.000)
		500	0.118 (0.000)	0.053 (0.000)	0.211 (0.000)	0.229 (0.001)	1.328 (0.002)	0.864 (0.000)	0.937 (0.000)
ZI-PGLD	CML	100	0.102 (0.001)	0.056 (0.000)	0.234 (0.000)	0.197 (0.000)	0.926 (0.000)	0.911 (0.000)	0.986 (0.001)
		300	0.089 (0.000)	0.049 (0.000)	0.222 (0.001)	0.206 (0.000)	0.889 (0.000)	0.864 (0.000)	0.996 (0.000)
		500	0.116 (0.000)	0.053 (0.000)	0.189 (0.000)	0.231 (0.000)	0.912 (0.000)	0.903 (0.000)	0.916 (0.000)
	CLS	100	0.096 (0.000)	0.052 (0.000)	0.196 (0.001)	0.199 (0.000)	0.896 (0.000)	0.925 (0.000)	0.943 (0.000)
		300	0.146 (0.000)	0.051 (0.000)	0.206 (0.000)	0.247 (0.000)	0.901 (0.000)	0.889 (0.000)	0.875 (0.000)
		500	0.138 (0.000)	0.046 (0.000)	0.228 (0.000)	0.229 (0.000)	0.934 (0.000)	0.943 (0.000)	0.923 (0.000)

TABLE 3.8

Simulated estimates and corresponding standard errors in () under different Zero-Inflated (ZI) innovations with binomial thinning, assuming stationarity

Innovation	Method	T	$\rho_1=0.1$	$\rho_2=0.05$	$\rho_3=0.2$	$\rho_4=0.3$	$\beta_0=0.3$	$\beta_1=0.2$	$\eta_0=0.3$	$\eta_1=0.2$	ν
ZI-Geometric	CML	100	0.097	0.051	0.232	0.346	0.302	0.274	0.264	0.294	
			(0.007)	(0.006)	(0.005)	(0.006)	(0.004)	(0.008)	(0.009)	(0.009)	
		300	0.112	0.047	0.196	0.331	0.323	0.223	0.302	0.212	
			(0.005)	(0.004)	(0.004)	(0.005)	(0.003)	(0.007)	(0.006)	(0.003)	
		500	0.126	0.044	0.214	0.322	0.341	0.196	0.296	0.231	
			(0.004)	(0.003)	(0.003)	(0.004)	(0.002)	(0.004)	(0.004)	(0.002)	
	CLS	100	0.122	0.049	0.198	0.317	0.289	0.196	0.286	0.196	
			(0.012)	(0.011)	(0.012)	(0.009)	(0.012)	(0.009)	(0.011)	(0.006)	
		300	0.142	0.053	0.203	0.288	0.311	0.234	0.309	0.231	
			(0.011)	(0.009)	(0.009)	(0.008)	(0.007)	(0.008)	(0.009)	(0.004)	
		500	0.099	0.051	0.188	0.326	0.299	0.203	0.249	0.211	
			(0.008)	(0.004)	(0.007)	(0.005)	(0.004)	(0.006)	(0.007)	(0.003)	
ZI-NB	CML	100	0.148	0.052	0.211	0.326	0.312	0.187	0.305	0.226	0.689
			(0.003)	(0.007)	(0.005)	(0.005)	(0.003)	(0.006)	(0.005)	(0.003)	(0.005)
		300	0.089	0.056	0.189	0.318	0.329	0.212	0.317	0.194	0.861
			(0.007)	(0.006)	(0.004)	(0.004)	(0.002)	(0.005)	(0.003)	(0.002)	(0.004)
		500	0.096	0.050	0.231	0.336	0.289	0.223	0.286	0.214	0.982
			(0.005)	(0.002)	(0.003)	(0.002)	(0.001)	(0.004)	(0.002)	(0.001)	(0.003)
	CLS	100	0.101	0.047	0.229	0.341	0.326	0.248	0.275	0.249	0.764
			(0.006)	(0.008)	(0.009)	(0.011)	(0.009)	(0.015)	(0.013)	(0.006)	(0.009)
		300	0.098	0.052	0.203	0.289	0.279	0.198	0.311	0.315	0.843
			(0.004)	(0.005)	(0.007)	(0.008)	(0.007)	(0.011)	(0.011)	(0.005)	(0.006)
		500	0.111	0.047	0.197	0.301	0.332	0.207	0.333	0.289	0.912
			(0.003)	(0.003)	(0.004)	(0.005)	(0.004)	(0.009)	(0.009)	(0.003)	(0.005)
ZI-CMP	CML	100	0.111	0.058	0.231	0.374	0.336	0.196	0.333	0.206	0.841
			(0.014)	(0.009)	(0.004)	(0.009)	(0.007)	(0.006)	(0.006)	(0.012)	(0.003)
		300	0.088	0.053	0.222	0.321	0.289	0.231	0.288	0.243	0.756
			(0.009)	(0.006)	(0.002)	(0.008)	(0.006)	(0.004)	(0.005)	(0.008)	(0.005)
		500	0.133	0.046	0.206	0.303	0.312	0.248	0.321	0.211	0.897
			(0.006)	(0.004)	(0.001)	(0.006)	(0.005)	(0.003)	(0.004)	(0.006)	(0.002)
	CLS	100	0.108	0.056	0.213	0.324	0.277	0.221	0.341	0.216	0.853
			(0.019)	(0.016)	(0.009)	(0.012)	(0.011)	(0.013)	(0.009)	(0.015)	(0.006)
		300	0.099	0.049	0.196	0.316	0.342	0.196	0.266	0.188	0.806
			(0.015)	(0.011)	(0.004)	(0.011)	(0.009)	(0.009)	(0.005)	(0.011)	(0.007)
		500	0.112	0.052	0.243	0.289	0.327	0.213	0.316	0.231	0.779
			(0.009)	(0.009)	(0.003)	(0.008)	(0.007)	(0.005)	(0.004)	(0.008)	(0.004)
ZI-PL	CML	100	0.152	0.055	0.248	0.316	0.346	0.202	0.331	0.236	
			(0.016)	(0.019)	(0.009)	(0.005)	(0.009)	(0.007)	(0.011)	(0.013)	
		300	0.123	0.050	0.197	0.339	0.303	0.197	0.312	0.228	
			(0.011)	(0.014)	(0.004)	(0.004)	(0.006)	(0.005)	(0.006)	(0.008)	
		500	0.095	0.047	0.233	0.310	0.289	0.230	0.324	0.241	
			(0.009)	(0.009)	(0.003)	(0.003)	(0.005)	(0.004)	(0.009)	(0.006)	
	CLS	100	0.111	0.044	0.216	0.346	0.346	0.229	0.306	0.190	
			(0.021)	(0.019)	(0.012)	(0.013)	(0.012)	(0.009)	(0.018)	(0.016)	
		300	0.098	0.059	0.197	0.276	0.289	0.196	0.288	0.201	
			(0.018)	(0.016)	(0.009)	(0.011)	(0.009)	(0.007)	(0.014)	(0.013)	
		500	0.103	0.051	0.246	0.305	0.312	0.216	0.312	0.197	
			(0.015)	(0.011)	(0.007)	(0.009)	(0.007)	(0.004)	(0.011)	(0.009)	

Innovation	Method	T	$\rho_1=0.1$	$\rho_2=0.05$	$\rho_3=0.2$	$\rho_4=0.3$	$\beta_0=0.3$	$\beta_1=0.2$	$\eta_0=0.3$	$\eta_1=0.2$	ν
ZI-PT	CML	100	0.119	0.039	0.245	0.342	0.323	0.214	0.342	0.229	1.336
			(0.011)	(0.015)	(0.008)	(0.006)	(0.005)	(0.004)	(0.008)	(0.008)	(0.006)
		300	0.103	0.046	0.186	0.335	0.316	0.222	0.344	0.242	1.452
			(0.005)	(0.008)	(0.007)	(0.007)	(0.004)	(0.003)	(0.006)	(0.004)	(0.004)
		500	0.137	0.053	0.213	0.311	0.328	0.203	0.311	0.231	1.118
			(0.004)	(0.006)	(0.005)	(0.004)	(0.006)	(0.002)	(0.005)	(0.003)	(0.003)
	CLS	100	0.076	0.061	0.224	0.326	0.326	0.237	0.329	0.196	1.256
			(0.019)	(0.021)	(0.013)	(0.012)	(0.009)	(0.006)	(0.010)	(0.011)	(0.012)
		300	0.103	0.059	0.199	0.246	0.296	0.178	0.286	0.229	1.421
			(0.015)	(0.019)	(0.011)	(0.007)	(0.007)	(0.004)	(0.008)	(0.009)	(0.009)
		500	0.096	0.052	0.233	0.289	0.331	0.221	0.311	0.187	1.334
			(0.011)	(0.011)	(0.009)	(0.005)	(0.005)	(0.002)	(0.006)	(0.008)	(0.007)
ZI-PIG	CML	100	0.099	0.061	0.284	0.332	0.365	0.246	0.326	0.241	
			(0.018)	(0.017)	(0.009)	(0.013)	(0.012)	(0.009)	(0.008)	(0.005)	
		300	0.101	0.055	0.241	0.287	0.336	0.189	0.314	0.196	
			(0.014)	(0.015)	(0.008)	(0.008)	(0.008)	(0.006)	(0.005)	(0.004)	
		500	0.121	0.052	0.223	0.300	0.329	0.199	0.296	0.232	
			(0.012)	(0.009)	(0.004)	(0.005)	(0.006)	(0.005)	(0.004)	(0.002)	
	CLS	100	0.086	0.048	0.241	0.342	0.288	0.227	0.333	0.246	
			(0.025)	(0.023)	(0.012)	(0.008)	(0.016)	(0.012)	(0.015)	(0.009)	
		300	0.097	0.053	0.196	0.276	0.312	0.199	0.299	0.166	
			(0.021)	(0.017)	(0.010)	(0.005)	(0.013)	(0.009)	(0.012)	(0.008)	
		500	0.122	0.062	0.231	0.309	0.299	0.213	0.346	0.229	
			(0.018)	(0.015)	(0.008)	(0.003)	(0.011)	(0.006)	(0.009)	(0.005)	
ZI-WCG	CML	100	0.096	0.045	0.236	0.331	0.306	0.226	0.312	0.229	0.986
			(0.007)	(0.004)	(0.006)	(0.003)	(0.005)	(0.002)	(0.011)	(0.004)	(0.005)
		300	0.147	0.052	0.196	0.309	0.319	0.196	0.334	0.196	1.027
			(0.006)	(0.003)	(0.003)	(0.002)	(0.003)	(0.001)	(0.005)	(0.003)	(0.004)
		500	0.122	0.056	0.206	0.296	0.286	0.219	0.326	0.206	1.322
			(0.004)	(0.003)	(0.002)	(0.001)	(0.002)	(0.000)	(0.003)	(0.001)	(0.003)
	CLS	100	0.123	0.048	0.236	0.312	0.344	0.244	0.327	0.188	1.102
			(0.011)	(0.009)	(0.008)	(0.007)	(0.009)	(0.005)	(0.013)	(0.006)	(0.009)
		300	0.097	0.052	0.199	0.289	0.287	0.196	0.266	0.228	1.023
			(0.009)	(0.007)	(0.006)	(0.005)	(0.006)	(0.004)	(0.011)	(0.005)	(0.006)
		500	0.112	0.049	0.223	0.346	0.308	0.213	0.311	0.199	1.226
			(0.007)	(0.005)	(0.005)	(0.003)	(0.004)	(0.002)	(0.009)	(0.003)	(0.005)
ZI-PWE	CML	100	0.113	0.051	0.204	0.311	0.352	0.227	0.333	0.199	0.855
			(0.002)	(0.003)	(0.000)	(0.002)	(0.001)	(0.002)	(0.001)	(0.000)	(0.002)
		300	0.096	0.046	0.196	0.296	0.296	0.196	0.341	0.208	0.889
			(0.001)	(0.002)	(0.000)	(0.001)	(0.000)	(0.001)	(0.000)	(0.000)	(0.001)
		500	0.146	0.053	0.221	0.348	0.301	0.246	0.296	0.236	0.902
			(0.000)	(0.001)	(0.000)	(0.000)	(0.000)	(0.001)	(0.000)	(0.001)	(0.000)
	CLS	100	0.112	0.046	0.227	0.321	0.342	0.212	0.301	0.198	0.889
			(0.003)	(0.005)	(0.004)	(0.003)	(0.003)	(0.004)	(0.002)	(0.003)	(0.004)
		300	0.097	0.063	0.188	0.287	0.298	0.156	0.288	0.202	0.941
			(0.002)	(0.003)	(0.003)	(0.002)	(0.002)	(0.002)	(0.001)	(0.001)	(0.003)
		500	0.105	0.052	0.211	0.305	0.323	0.206	0.344	0.179	0.903
			(0.001)	(0.002)	(0.002)	(0.000)	(0.001)	(0.001)	(0.000)	(0.002)	(0.002)
ZI-PGLD	CML	100	0.149	0.052	0.237	0.310	0.255	0.199	0.319	0.236	0.943
			(0.001)	(0.000)	(0.001)	(0.000)	(0.000)	(0.000)	(0.000)	(0.001)	(0.000)
		300	0.100	0.048	0.198	0.289	0.326	0.211	0.311	0.189	0.896
			(0.001)	(0.000)	(0.000)	(0.000)	(0.000)	(0.000)	(0.000)	(0.000)	(0.000)
		500	0.112	0.056	0.204	0.306	0.297	0.238	0.259	0.206	0.967
			(0.000)	(0.000)	(0.000)	(0.000)	(0.000)	(0.000)	(0.000)	(0.000)	(0.000)
	CLS	100	0.103	0.052	0.156	0.256	0.328	0.196	0.326	0.216	0.980
			(0.000)	(0.000)	(0.000)	(0.000)	(0.000)	(0.000)	(0.000)	(0.000)	(0.000)
		300	0.089	0.045	0.234	0.344	0.289	0.228	0.299	0.195	0.956
			(0.001)	(0.000)	(0.000)	(0.000)	(0.000)	(0.000)	(0.000)	(0.000)	(0.000)
		500	0.136	0.053	0.179	0.299	0.309	0.177	0.306	0.188	0.897
			(0.000)	(0.000)	(0.000)	(0.000)	(0.000)	(0.000)	(0.000)	(0.000)	(0.000)

TABLE 3.9

Simulated estimates and corresponding standard errors in () under different Zero-Inflated (ZI) innovations with binomial thinning, assuming non-stationarity

Innovation	Method	T	$\rho_1=0.1$	$\rho_2=0.05$	$\rho_3=0.2$	$\rho_4=0.3$	λ_t	γ
Poisson	CML	100	0.132 (0.001)	0.043 (0.002)	0.212 (0.003)	0.325 (0.003)	0.865 (0.002)	0.896 (0.003)
		300	0.096 (0.000)	0.052 (0.000)	0.223 (0.002)	0.265 (0.001)	0.912 (0.001)	0.996 (0.002)
		500	0.112 (0.000)	0.056 (0.001)	0.196 (0.001)	0.301 (0.000)	0.897 (0.000)	0.852 (0.001)
	CLS	100	0.158 (0.003)	0.053 (0.004)	0.189 (0.004)	0.319 (0.004)	0.923 (0.003)	0.965 (0.004)
		300	0.164 (0.002)	0.049 (0.002)	0.223 (0.002)	0.286 (0.002)	0.974 (0.002)	0.988 (0.003)
		500	0.112 (0.001)	0.057 (0.001)	0.235 (0.000)	0.320 (0.000)	0.855 (0.002)	0.971 (0.002)

TABLE 3.10
Simulated estimates and corresponding standard errors in () under Poisson innovation with Generalized Binomial thinning, assuming stationary setting

Innovation	Method	T	$\rho_1=0.1$	$\rho_2=0.05$	$\rho_3=0.2$	$\rho_4=0.3$	$\beta_0=0.3$	$\beta_1=0.2$	γ
Poisson	CML	100	0.144 (0.004)	0.048 (0.004)	0.222 (0.003)	0.323 (0.005)	0.286 (0.002)	0.228 (0.003)	0.877 (0.005)
		300	0.119 (0.003)	0.051 (0.003)	0.237 (0.002)	0.298 (0.003)	0.327 (0.001)	0.199 (0.002)	0.869 (0.004)
		500	0.099 (0.002)	0.046 (0.002)	0.212 (0.001)	0.332 (0.002)	0.300 (0.000)	0.201 (0.001)	0.926 (0.003)
	CLS	100	0.111 (0.006)	0.057 (0.007)	0.196 (0.005)	0.346 (0.006)	0.266 (0.005)	0.189 (0.007)	0.861 (0.004)
		300	0.141 (0.005)	0.052 (0.009)	0.187 (0.003)	0.269 (0.004)	0.283 (0.004)	0.249 (0.008)	0.923 (0.003)
		500	0.106 (0.004)	0.051 (0.004)	0.202 (0.002)	0.307 (0.002)	0.311 (0.002)	0.163 (0.006)	0.908 (0.002)

TABLE 3.11
Simulated estimates and corresponding standard errors in () under Poisson innovation with Generalized Binomial thinning, assuming non-stationary setting

Innovation	Method	T	$\rho_1=0.1$	$\rho_2=0.05$	$\rho_3=0.2$	$\rho_4=0.3$	λ_t	π_t	γ
ZI-P	CML	100	0.126 (0.004)	0.052 (0.006)	0.185 (0.003)	0.305 (0.004)	0.876 (0.003)	0.910 (0.006)	0.871 (0.011)
		300	0.131 (0.002)	0.055 (0.004)	0.176 (0.004)	0.286 (0.003)	0.905 (0.001)	0.869 (0.004)	0.962 (0.006)
		500	0.095 (0.001)	0.053 (0.002)	0.203 (0.001)	0.311 (0.002)	0.916 (0.000)	0.921 (0.002)	0.996 (0.004)
	CLS	100	0.142 (0.006)	0.060 (0.008)	0.215 (0.005)	0.348 (0.007)	0.889 (0.005)	0.906 (0.008)	0.931 (0.013)
		300	0.104 (0.002)	0.046 (0.005)	0.196 (0.004)	0.286 (0.006)	0.896 (0.003)	0.855 (0.006)	0.887 (0.008)
		500	0.121 (0.001)	0.051 (0.003)	0.211 (0.003)	0.299 (0.003)	0.927 (0.001)	0.902 (0.005)	0.869 (0.007)

TABLE 3.12
Simulated estimates and corresponding standard errors in () under ZI-P innovation with Generalized Binomial thinning, assuming stationary setting

Innovation	Method	T	$\rho_1=0.1$	$\rho_2=0.05$	$\rho_3=0.2$	$\rho_4=0.3$	$\beta_0=0.3$	$\beta_1=0.2$	$\eta_0=0.3$	$\eta_1=0.2$	γ
ZI-P	CML	100	0.112 (0.006)	0.050 (0.005)	0.223 (0.005)	0.319 (0.007)	0.259 (0.008)	0.261 (0.004)	0.314 (0.006)	0.202 (0.004)	0.954 (0.006)
		300	0.103 (0.005)	0.046 (0.004)	0.186 (0.004)	0.286 (0.005)	0.328 (0.006)	0.234 (0.003)	0.333 (0.005)	0.195 (0.002)	0.998 (0.004)
		500	0.142 (0.002)	0.051 (0.002)	0.246 (0.002)	0.252 (0.003)	0.342 (0.004)	0.177 (0.002)	0.268 (0.002)	0.219 (0.002)	0.957 (0.003)
	CLS	100	0.088 (0.009)	0.055 (0.008)	0.169 (0.010)	0.327 (0.008)	0.339 (0.007)	0.237 (0.006)	0.289 (0.008)	0.188 (0.004)	0.945 (0.011)
		300	0.091 (0.006)	0.047 (0.006)	0.234 (0.006)	0.311 (0.009)	0.266 (0.006)	0.255 (0.005)	0.324 (0.005)	0.216 (0.003)	0.889 (0.009)
		500	0.133 (0.005)	0.056 (0.005)	0.225 (0.004)	0.329 (0.005)	0.282 (0.003)	0.218 (0.004)	0.345 (0.004)	0.222 (0.002)	0.994 (0.005)

TABLE 3.13
Simulated estimates and corresponding standard errors in () under ZI-P innovation with Generalized Binomial thinning, assuming non-stationary setting

Innovation	Method	T	$\rho_1=0.1$	$\rho_2=0.05$	$\rho_3=0.2$	$\rho_4=0.3$	λ_t	ν	γ
Geometric	CML	100	0.126	0.051	0.222	0.312	0.896		0.763
			(0.008)	(0.005)	(0.006)	(0.004)	(0.003)		(0.004)
		300	0.089	0.050	0.187	0.286	0.901		0.826
			(0.003)	(0.004)	(0.005)	(0.002)	(0.002)		(0.003)
		500	0.113	0.047	0.216	0.327	0.916		0.931
			(0.002)	(0.003)	(0.004)	(0.001)	(0.001)		(0.002)
	CLS	100	0.089	0.057	0.243	0.346	0.943		0.875
			(0.009)	(0.010)	(0.007)	(0.008)	(0.007)		(0.012)
		300	0.123	0.048	0.196	0.286	0.876		0.905
			(0.008)	(0.008)	(0.005)	(0.005)	(0.006)		(0.009)
		500	0.106	0.053	0.201	0.300	0.926		0.846
			(0.005)	(0.007)	(0.004)	(0.003)	(0.005)		(0.008)
NB	CML	100	0.125	0.053	0.166	0.299	0.886	0.754	0.866
			(0.002)	(0.003)	(0.004)	(0.003)	(0.003)	(0.001)	(0.003)
		300	0.147	0.047	0.232	0.311	0.875	0.862	0.917
			(0.001)	(0.002)	(0.003)	(0.002)	(0.001)	(0.000)	(0.001)
		500	0.120	0.056	0.214	0.326	0.924	0.965	0.862
			(0.000)	(0.001)	(0.001)	(0.001)	(0.000)	(0.000)	(0.000)
	CLS	100	0.111	0.059	0.218	0.309	0.946	0.817	0.868
			(0.006)	(0.005)	(0.009)	(0.008)	(0.004)	(0.003)	(0.009)
		300	0.096	0.048	0.166	0.275	0.887	0.789	0.778
			(0.004)	(0.004)	(0.007)	(0.005)	(0.003)	(0.002)	(0.005)
		500	0.142	0.051	0.226	0.312	0.943	0.806	0.922
			(0.003)	(0.003)	(0.005)	(0.003)	(0.002)	(0.001)	(0.003)
CMP	CML	100	0.097	0.059	0.196	0.316	1.226	0.823	0.916
			(0.006)	(0.009)	(0.009)	(0.012)	(0.008)	(0.012)	(0.005)
		300	0.102	0.061	0.235	0.352	1.024	0.814	0.987
			(0.004)	(0.007)	(0.006)	(0.008)	(0.006)	(0.009)	(0.004)
		500	0.112	0.051	0.194	0.265	1.462	0.759	0.773
			(0.003)	(0.005)	(0.004)	(0.006)	(0.003)	(0.007)	(0.003)
	CLS	100	0.131	0.053	0.241	0.322	1.453	0.942	0.865
			(0.013)	(0.011)	(0.015)	(0.009)	(0.013)	(0.019)	(0.006)
		300	0.098	0.049	0.198	0.290	1.126	0.889	0.908
			(0.009)	(0.007)	(0.009)	(0.008)	(0.009)	(0.015)	(0.005)
		500	0.123	0.051	0.244	0.302	1.327	0.906	0.876
			(0.008)	(0.005)	(0.007)	(0.007)	(0.007)	(0.009)	(0.003)
PL	CML	100	0.127	0.046	0.216	0.258	1.013		0.822
			(0.011)	(0.009)	(0.008)	(0.006)	(0.004)		(0.006)
		300	0.113	0.053	0.225	0.302	1.236		0.914
			(0.005)	(0.007)	(0.003)	(0.005)	(0.003)		(0.002)
		500	0.107	0.058	0.247	0.297	1.401		0.933
			(0.004)	(0.005)	(0.002)	(0.004)	(0.002)		(0.001)
	CLS	100	0.134	0.061	0.213	0.298	1.122		0.927
			(0.016)	(0.013)	(0.009)	(0.014)	(0.009)		(0.009)
		300	0.075	0.059	0.196	0.316	1.036		0.865
			(0.019)	(0.009)	(0.008)	(0.009)	(0.007)		(0.008)
		500	0.103	0.043	0.203	0.346	1.464		0.943
			(0.012)	(0.005)	(0.006)	(0.007)	(0.005)		(0.007)

Innovation	Method	T	$\rho_1=0.1$	$\rho_2=0.05$	$\rho_3=0.2$	$\rho_4=0.3$	λ_t	v	γ
PT	CML	100	0.122 (0.012)	0.046 (0.008)	0.196 (0.005)	0.311 (0.006)	0.906 (0.009)	1.232 (0.002)	0.876 (0.003)
		300	0.118 (0.009)	0.049 (0.006)	0.207 (0.003)	0.286 (0.004)	0.886 (0.007)	1.304 (0.001)	0.937 (0.002)
		500	0.103 (0.008)	0.052 (0.003)	0.212 (0.001)	0.305 (0.002)	0.923 (0.005)	1.678 (0.000)	0.886 (0.001)
	CLS	100	0.098 (0.019)	0.046 (0.011)	0.233 (0.010)	0.278 (0.008)	0.886 (0.012)	1.445 (0.009)	0.942 (0.011)
		300	0.103 (0.017)	0.053 (0.007)	0.198 (0.009)	0.326 (0.007)	0.904 (0.009)	1.621 (0.007)	0.856 (0.008)
		500	0.117 (0.011)	0.066 (0.009)	0.203 (0.007)	0.296 (0.005)	0.892 (0.007)	1.023 (0.005)	0.902 (0.007)
PIG	CML	100	0.114 (0.019)	0.058 (0.015)	0.187 (0.007)	0.312 (0.005)	0.889 (0.003)		0.916 (0.005)
		300	0.089 (0.017)	0.047 (0.011)	0.223 (0.005)	0.352 (0.003)	0.911 (0.002)		0.756 (0.004)
		500	0.100 (0.016)	0.052 (0.008)	0.196 (0.003)	0.286 (0.002)	0.932 (0.001)		0.811 (0.003)
	CLS	100	0.098 (0.025)	0.053 (0.019)	0.216 (0.009)	0.316 (0.009)	0.923 (0.006)		0.886 (0.009)
		300	0.123 (0.019)	0.049 (0.013)	0.196 (0.007)	0.289 (0.007)	0.886 (0.004)		0.926 (0.008)
		500	0.119 (0.015)	0.055 (0.009)	0.241 (0.005)	0.303 (0.005)	0.903 (0.002)		0.866 (0.005)
WCG	CML	100	0.121 (0.003)	0.051 (0.003)	0.202 (0.002)	0.317 (0.001)	1.041 (0.006)	1.232 (0.002)	0.763 (0.002)
		300	0.096 (0.001)	0.048 (0.002)	0.231 (0.001)	0.322 (0.000)	1.871 (0.004)	1.356 (0.001)	0.974 (0.001)
		500	0.134 (0.000)	0.054 (0.001)	0.198 (0.000)	0.341 (0.000)	1.232 (0.003)	1.127 (0.000)	0.771 (0.000)
	CLS	100	0.106 (0.008)	0.048 (0.005)	0.231 (0.006)	0.326 (0.004)	1.056 (0.009)	1.021 (0.005)	0.935 (0.004)
		300	0.089 (0.007)	0.053 (0.003)	0.199 (0.003)	0.296 (0.003)	1.456 (0.006)	1.426 (0.003)	0.865 (0.003)
		500	0.136 (0.005)	0.049 (0.002)	0.202 (0.001)	0.312 (0.002)	1.235 (0.005)	1.211 (0.002)	0.922 (0.002)
PWE	CML	100	0.126 (0.002)	0.055 (0.000)	0.188 (0.001)	0.306 (0.000)	1.112 (0.000)	0.852 (0.000)	0.842 (0.002)
		300	0.108 (0.001)	0.047 (0.000)	0.226 (0.000)	0.286 (0.000)	1.235 (0.000)	0.814 (0.000)	0.789 (0.001)
		500	0.144 (0.000)	0.046 (0.000)	0.247 (0.000)	0.311 (0.000)	1.449 (0.000)	0.899 (0.000)	0.932 (0.000)
	CLS	100	0.096 (0.002)	0.050 (0.000)	0.174 (0.003)	0.318 (0.003)	1.096 (0.002)	0.859 (0.001)	0.903 (0.003)
		300	0.102 (0.000)	0.049 (0.001)	0.236 (0.001)	0.287 (0.002)	1.854 (0.001)	0.842 (0.000)	0.889 (0.002)
		500	0.115 (0.000)	0.052 (0.000)	0.196 (0.000)	0.311 (0.000)	1.754 (0.000)	0.777 (0.000)	0.941 (0.000)
PGLD	CML	100	0.131 (0.000)	0.052 (0.000)	0.214 (0.000)	0.323 (0.000)	0.956 (0.000)	0.981 (0.000)	0.966 (0.000)
		300	0.116 (0.000)	0.048 (0.000)	0.156 (0.000)	0.296 (0.000)	0.865 (0.000)	0.956 (0.000)	0.968 (0.000)
		500	0.124 (0.000)	0.053 (0.000)	0.208 (0.000)	0.301 (0.000)	0.978 (0.000)	0.924 (0.000)	0.865 (0.000)
	CLS	100	0.103 (0.001)	0.046 (0.001)	0.220 (0.000)	0.289 (0.000)	0.996 (0.002)	0.916 (0.000)	0.859 (0.000)
		300	0.111 (0.000)	0.045 (0.001)	0.189 (0.000)	0.316 (0.000)	0.879 (0.000)	0.965 (0.000)	0.926 (0.000)
		500	0.096 (0.000)	0.056 (0.000)	0.216 (0.000)	0.325 (0.000)	0.889 (0.000)	1.021 (0.000)	0.894 (0.000)

TABLE 3.14

Simulated estimates and corresponding standard errors in () under different Poisson-mixture innovations with Generalized Binomial thinning, assuming stationarity

Innovation	Method	T	$\rho_1=0.1$	$\rho_2=0.05$	$\rho_3=0.2$	$\rho_4=0.3$	$\beta_0=0.3$	$\beta_1=0.2$	v	γ
Geometric	CML	100	0.115	0.056	0.239	0.289	0.347	0.224		0.852
			(0.009)	(0.011)	(0.006)	(0.004)	(0.007)	(0.006)		(0.005)
		300	0.097	0.049	0.251	0.302	0.296	0.196		0.936
			(0.008)	(0.008)	(0.005)	(0.002)	(0.005)	(0.003)		(0.003)
		500	0.141	0.053	0.198	0.319	0.321	0.246		0.889
			(0.006)	(0.006)	(0.004)	(0.000)	(0.003)	(0.004)		(0.001)
	CLS	100	0.104	0.046	0.196	0.301	0.290	0.200		0.922
			(0.013)	(0.009)	(0.009)	(0.006)	(0.011)	(0.005)		(0.009)
		300	0.096	0.053	0.203	0.326	0.312	0.246		0.886
			(0.009)	(0.008)	(0.007)	(0.005)	(0.009)	(0.004)		(0.008)
		500	0.111	0.047	0.188	0.288	0.279	0.196		0.926
			(0.005)	(0.005)	(0.005)	(0.004)	(0.005)	(0.003)		(0.005)
NB	CML	100	0.095	0.039	0.216	0.324	0.315	0.243	0.777	0.902
			(0.015)	(0.006)	(0.011)	(0.005)	(0.008)	(0.003)	(0.004)	(0.003)
		300	0.102	0.047	0.186	0.276	0.289	0.188	0.816	0.890
			(0.012)	(0.004)	(0.008)	(0.003)	(0.009)	(0.002)	(0.003)	(0.002)
		500	0.116	0.051	0.203	0.349	0.296	0.176	0.947	0.832
			(0.009)	(0.003)	(0.005)	(0.002)	(0.007)	(0.001)	(0.002)	(0.000)
	CLS	100	0.076	0.063	0.187	0.346	0.321	0.221	0.943	0.796
			(0.020)	(0.009)	(0.016)	(0.011)	(0.015)	(0.005)	(0.008)	(0.009)
		300	0.139	0.061	0.226	0.286	0.289	0.198	0.889	0.866
			(0.018)	(0.008)	(0.012)	(0.009)	(0.011)	(0.004)	(0.007)	(0.008)
		500	0.141	0.059	0.198	0.199	0.306	0.203	0.913	0.911
			(0.014)	(0.007)	(0.009)	(0.007)	(0.009)	(0.003)	(0.005)	(0.007)
CMP	CML	100	0.122	0.042	0.198	0.341	0.311	0.219	0.798	0.821
			(0.019)	(0.010)	(0.006)	(0.008)	(0.009)	(0.006)	(0.010)	(0.006)
		300	0.089	0.046	0.238	0.329	0.349	0.185	0.865	0.769
			(0.016)	(0.009)	(0.005)	(0.005)	(0.007)	(0.005)	(0.006)	(0.004)
		500	0.143	0.050	0.179	0.296	0.331	0.232	0.802	0.926
			(0.011)	(0.006)	(0.004)	(0.004)	(0.006)	(0.003)	(0.004)	(0.003)
	CLS	100	0.089	0.036	0.213	0.306	0.277	0.222	0.942	0.933
			(0.025)	(0.013)	(0.009)	(0.011)	(0.008)	(0.009)	(0.012)	(0.009)
		300	0.097	0.044	0.199	0.279	0.316	0.196	0.879	0.886
			(0.021)	(0.009)	(0.008)	(0.009)	(0.004)	(0.008)	(0.009)	(0.008)
		500	0.146	0.055	0.246	0.200	0.282	0.241	0.903	0.912
			(0.019)	(0.007)	(0.007)	(0.007)	(0.005)	(0.005)	(0.008)	(0.005)
PL	CML	100	0.087	0.048	0.223	0.316	0.298	0.230		0.923
			(0.019)	(0.015)	(0.008)	(0.011)	(0.009)	(0.006)		(0.008)
		300	0.096	0.053	0.189	0.286	0.327	0.221		0.811
			(0.019)	(0.014)	(0.006)	(0.009)	(0.008)	(0.004)		(0.005)
		500	0.152	0.056	0.214	0.303	0.254	0.243		0.917
			(0.017)	(0.013)	(0.005)	(0.006)	(0.006)	(0.003)		(0.003)
	CLS	100	0.156	0.055	0.231	0.279	0.302	0.246		0.889
			(0.031)	(0.019)	(0.011)	(0.009)	(0.015)	(0.012)		(0.007)
		300	0.144	0.043	0.197	0.312	0.279	0.187		0.912
			(0.022)	(0.015)	(0.008)	(0.008)	(0.009)	(0.009)		(0.005)
		500	0.096	0.054	0.224	0.256	0.312	0.196		0.946
			(0.019)	(0.011)	(0.005)	(0.007)	(0.007)	(0.005)		(0.003)

Innovation	Method	T	$\rho_1=0.1$	$\rho_2=0.05$	$\rho_3=0.2$	$\rho_4=0.3$	$\beta_0=0.3$	$\beta_1=0.2$	ν	γ
PT	CML	100	0.086 (0.017)	0.038 (0.016)	0.235 (0.015)	0.318 (0.012)	0.289 (0.018)	0.216 (0.011)	1.030 (0.007)	0.862 (0.009)
		300	0.101 (0.016)	0.041 (0.014)	0.219 (0.013)	0.288 (0.008)	0.175 (0.016)	0.229 (0.009)	1.264 (0.005)	0.902 (0.006)
		500	0.123 (0.014)	0.046 (0.012)	0.198 (0.011)	0.329 (0.003)	0.231 (0.012)	0.237 (0.004)	1.454 (0.003)	0.821 (0.005)
	CLS	100	0.155 (0.021)	0.066 (0.017)	0.196 (0.014)	0.346 (0.015)	0.256 (0.025)	0.199 (0.012)	1.326 (0.028)	0.889 (0.014)
		300	0.078 (0.018)	0.044 (0.015)	0.232 (0.013)	0.317 (0.008)	0.326 (0.015)	0.222 (0.009)	1.678 (0.021)	0.789 (0.015)
		500	0.149 (0.016)	0.059 (0.011)	0.246 (0.009)	0.269 (0.007)	0.341 (0.009)	0.216 (0.008)	1.023 (0.019)	0.856 (0.012)
PIG	CML	100	0.161 (0.015)	0.051 (0.010)	0.201 (0.012)	0.322 (0.009)	0.296 (0.015)	0.203 (0.011)		0.773 (0.017)
		300	0.132 (0.012)	0.049 (0.006)	0.199 (0.009)	0.346 (0.007)	0.315 (0.009)	0.196 (0.009)		0.825 (0.015)
		500	0.111 (0.009)	0.045 (0.004)	0.232 (0.004)	0.266 (0.003)	0.267 (0.007)	0.231 (0.006)		0.923 (0.009)
	CLS	100	0.126 (0.017)	0.066 (0.016)	0.189 (0.015)	0.289 (0.011)	0.309 (0.019)	0.189 (0.015)		0.816 (0.019)
		300	0.089 (0.015)	0.058 (0.013)	0.231 (0.009)	0.341 (0.009)	0.289 (0.015)	0.249 (0.012)		0.789 (0.016)
		500	0.149 (0.011)	0.054 (0.012)	0.217 (0.005)	0.303 (0.005)	0.316 (0.007)	0.175 (0.009)		0.941 (0.012)
WCG	CML	100	0.142 (0.006)	0.056 (0.008)	0.211 (0.007)	0.331 (0.006)	0.289 (0.005)	0.241 (0.009)	0.883 (0.006)	0.941 (0.006)
		300	0.125 (0.003)	0.062 (0.006)	0.156 (0.003)	0.267 (0.004)	0.326 (0.003)	0.238 (0.008)	1.026 (0.005)	0.882 (0.005)
		500	0.132 (0.001)	0.044 (0.004)	0.175 (0.002)	0.312 (0.003)	0.304 (0.002)	0.168 (0.006)	1.325 (0.004)	0.752 (0.003)
	CLS	100	0.112 (0.009)	0.058 (0.014)	0.209 (0.009)	0.349 (0.013)	0.296 (0.009)	0.188 (0.012)	0.665 (0.012)	0.896 (0.009)
		300	0.146 (0.008)	0.055 (0.009)	0.189 (0.008)	0.288 (0.008)	0.311 (0.006)	0.256 (0.009)	1.022 (0.010)	0.936 (0.008)
		500	0.098 (0.005)	0.041 (0.005)	0.246 (0.007)	0.311 (0.007)	0.296 (0.005)	0.189 (0.008)	1.364 (0.009)	0.911 (0.006)
PWE	CML	100	0.127 (0.000)	0.053 (0.000)	0.168 (0.001)	0.326 (0.002)	0.331 (0.000)	0.157 (0.002)	0.912 (0.001)	0.817 (0.003)
		300	0.106 (0.000)	0.050 (0.000)	0.199 (0.000)	0.289 (0.001)	0.266 (0.000)	0.208 (0.000)	0.946 (0.000)	0.923 (0.002)
		500	0.119 (0.000)	0.047 (0.000)	0.203 (0.000)	0.304 (0.000)	0.278 (0.000)	0.186 (0.000)	0.859 (0.000)	0.746 (0.001)
	CLS	100	0.098 (0.000)	0.049 (0.002)	0.222 (0.000)	0.349 (0.002)	0.289 (0.001)	0.211 (0.004)	0.914 (0.003)	0.888 (0.005)
		300	0.123 (0.000)	0.051 (0.001)	0.199 (0.000)	0.286 (0.001)	0.302 (0.000)	0.176 (0.002)	0.876 (0.002)	0.921 (0.003)
		500	0.144 (0.000)	0.054 (0.000)	0.243 (0.000)	0.322 (0.000)	0.343 (0.000)	0.234 (0.000)	0.941 (0.001)	0.937 (0.002)
PGLD	CML	100	0.112 (0.001)	0.050 (0.000)	0.231 (0.000)	0.340 (0.000)	0.336 (0.000)	0.199 (0.000)	0.956 (0.000)	0.889 (0.001)
		300	0.118 (0.000)	0.053 (0.000)	0.219 (0.000)	0.317 (0.000)	0.312 (0.000)	0.236 (0.000)	0.989 (0.000)	1.023 (0.000)
		500	0.104 (0.000)	0.048 (0.000)	0.232 (0.000)	0.328 (0.000)	0.256 (0.000)	0.156 (0.000)	0.896 (0.000)	0.965 (0.000)
	CLS	100	0.121 (0.000)	0.051 (0.000)	0.221 (0.000)	0.256 (0.000)	0.305 (0.000)	0.238 (0.000)	0.979 (0.000)	0.994 (0.000)
		300	0.126 (0.000)	0.050 (0.000)	0.241 (0.000)	0.312 (0.000)	0.346 (0.000)	0.206 (0.000)	0.889 (0.000)	0.996 (0.000)
		500	0.107 (0.000)	0.045 (0.000)	0.233 (0.000)	0.289 (0.000)	0.256 (0.000)	0.244 (0.000)	0.945 (0.000)	0.865 (0.000)

TABLE 3.15

Simulated estimates and corresponding standard errors in () under different Poisson-mixture innovations with Generalized Binomial thinning, assuming non-stationarity

Innovation	Method	T	$\rho_1=0.1$	$\rho_2=0.05$	$\rho_3=0.2$	$\rho_4=0.3$	λ_t	$\pi_t=0.9$	ν	γ
ZI-Geometric	CML	100	0.142	0.032	0.229	0.296	0.886	0.856		0.814
			(0.009)	(0.011)	(0.006)	(0.009)	(0.007)	(0.006)		(0.008)
		300	0.133	0.056	0.186	0.322	0.913	0.919		0.952
			(0.008)	(0.008)	(0.004)	(0.006)	(0.006)	(0.005)		(0.006)
		500	0.106	0.052	0.231	0.254	0.904	0.923		0.866
			(0.006)	(0.006)	(0.003)	(0.005)	(0.004)	(0.002)		(0.005)
	CLS	100	0.111	0.059	0.198	0.323	0.922	0.931		0.926
			(0.012)	(0.015)	(0.009)	(0.012)	(0.009)	(0.008)		(0.011)
		300	0.096	0.042	0.217	0.276	0.886	0.852		0.889
			(0.008)	(0.011)	(0.007)	(0.009)	(0.007)	(0.007)		(0.009)
		500	0.089	0.056	0.236	0.316	0.909	0.878		0.933
			(0.009)	(0.009)	(0.005)	(0.007)	(0.005)	(0.005)		(0.008)
ZI-NB	CML	100	0.119	0.049	0.196	0.305	0.875	0.899	0.861	0.906
			(0.009)	(0.010)	(0.008)	(0.006)	(0.007)	(0.003)	(0.002)	(0.004)
		300	0.136	0.044	0.201	0.321	0.932	0.936	0.786	0.827
			(0.006)	(0.008)	(0.007)	(0.005)	(0.006)	(0.002)	(0.001)	(0.003)
		500	0.146	0.055	0.195	0.281	0.941	0.866	0.897	0.776
			(0.005)	(0.006)	(0.005)	(0.003)	(0.004)	(0.001)	(0.000)	(0.001)
	CLS	100	0.089	0.064	0.232	0.324	0.878	0.855	0.823	0.859
			(0.012)	(0.009)	(0.009)	(0.008)	(0.009)	(0.005)	(0.004)	(0.005)
		300	0.108	0.059	0.187	0.289	0.889	0.923	0.921	0.916
			(0.009)	(0.008)	(0.007)	(0.007)	(0.008)	(0.004)	(0.003)	(0.004)
		500	0.126	0.054	0.206	0.326	0.902	0.896	0.867	0.889
			(0.007)	(0.004)	(0.005)	(0.005)	(0.007)	(0.003)	(0.002)	(0.003)
ZI-CMP	CML	100	0.102	0.056	0.235	0.300	1.125	0.921	0.789	0.952
			(0.009)	(0.012)	(0.016)	(0.008)	(0.006)	(0.009)	(0.008)	(0.006)
		300	0.095	0.052	0.249	0.314	1.423	0.877	0.816	0.869
			(0.006)	(0.009)	(0.011)	(0.006)	(0.005)	(0.007)	(0.006)	(0.004)
		500	0.087	0.046	0.223	0.266	1.765	0.905	0.899	0.963
			(0.005)	(0.006)	(0.009)	(0.004)	(0.003)	(0.005)	(0.005)	(0.003)
	CLS	100	0.098	0.049	0.186	0.302	1.003	0.789	0.813	0.921
			(0.011)	(0.015)	(0.017)	(0.009)	(0.009)	(0.013)	(0.009)	(0.009)
		300	0.106	0.061	0.237	0.315	1.216	0.856	0.778	0.889
			(0.009)	(0.016)	(0.015)	(0.007)	(0.008)	(0.012)	(0.005)	(0.007)
		500	0.122	0.055	0.222	0.267	1.456	0.899	0.811	0.906
			(0.008)	(0.014)	(0.011)	(0.006)	(0.007)	(0.008)	(0.004)	(0.005)
ZI-PL	CML	100	0.102	0.042	0.188	0.276	1.322	0.864		0.872
			(0.021)	(0.016)	(0.015)	(0.011)	(0.006)	(0.004)		(0.003)
		300	0.145	0.048	0.176	0.341	1.255	0.932		0.942
			(0.019)	(0.015)	(0.012)	(0.008)	(0.005)	(0.003)		(0.002)
		500	0.112	0.051	0.242	0.275	1.586	0.875		0.936
			(0.016)	(0.013)	(0.008)	(0.005)	(0.003)	(0.002)		(0.001)
	CLS	100	0.067	0.066	0.229	0.321	1.623	0.906		0.911
			(0.025)	(0.016)	(0.012)	(0.014)	(0.011)	(0.008)		(0.005)
		300	0.088	0.045	0.168	0.289	1.246	0.913		0.865
			(0.024)	(0.013)	(0.009)	(0.009)	(0.009)	(0.007)		(0.004)
		500	0.123	0.056	0.204	0.203	1.003	0.865		0.905
			(0.018)	(0.012)	(0.008)	(0.005)	(0.007)	(0.005)		(0.003)

Innovation	Method	T	$\rho_1=0.1$	$\rho_2=0.05$	$\rho_3=0.2$	$\rho_4=0.3$	λ_t	$\pi_t=0.9$	ν	γ
ZI-PT	CML	100	0.089	0.059	0.234	0.289	0.868	0.926	1.236	0.881
			(0.019)	(0.013)	(0.012)	(0.014)	(0.018)	(0.015)	(0.009)	(0.005)
		300	0.143	0.062	0.184	0.334	0.923	0.786	1.013	0.888
			(0.013)	(0.009)	(0.009)	(0.011)	(0.015)	(0.009)	(0.005)	(0.004)
		500	0.126	0.058	0.192	0.301	0.911	0.911	1.624	0.946
			(0.009)	(0.007)	(0.008)	(0.008)	(0.013)	(0.007)	(0.004)	(0.003)
	CLS	100	0.098	0.067	0.214	0.327	0.942	0.915	1.796	0.951
			(0.026)	(0.018)	(0.015)	(0.015)	(0.021)	(0.019)	(0.012)	(0.009)
		300	0.115	0.056	0.223	0.275	0.933	0.826	1.454	0.877
			(0.021)	(0.015)	(0.011)	(0.013)	(0.018)	(0.017)	(0.009)	(0.008)
		500	0.088	0.057	0.188	0.326	0.859	0.902	1.236	0.965
			(0.017)	(0.014)	(0.009)	(0.009)	(0.016)	(0.014)	(0.007)	(0.007)
ZI-PIG	CML	100	0.123	0.051	0.227	0.288	0.925	0.942		0.705
			(0.014)	(0.007)	(0.011)	(0.015)	(0.021)	(0.019)		(0.006)
		300	0.131	0.053	0.166	0.320	0.943	0.886		0.812
			(0.012)	(0.005)	(0.009)	(0.013)	(0.018)	(0.017)		(0.004)
		500	0.098	0.047	0.213	0.268	0.896	0.937		0.927
			(0.010)	(0.004)	(0.006)	(0.011)	(0.016)	(0.015)		(0.003)
	CLS	100	0.096	0.046	0.226	0.299	0.906	0.911		0.859
			(0.018)	(0.009)	(0.016)	(0.019)	(0.026)	(0.023)		(0.009)
		300	0.099	0.055	0.200	0.344	0.889	0.937		0.904
			(0.015)	(0.008)	(0.012)	(0.016)	(0.021)	(0.018)		(0.007)
		500	0.116	0.054	0.196	0.326	0.913	0.897		0.877
			(0.009)	(0.006)	(0.009)	(0.014)	(0.018)	(0.014)		(0.005)
ZI-WCG	CML	100	0.086	0.045	0.208	0.317	1.121	0.866	0.877	0.828
			(0.008)	(0.009)	(0.006)	(0.007)	(0.009)	(0.005)	(0.003)	(0.004)
		300	0.127	0.032	0.200	0.346	1.346	0.853	1.326	0.933
			(0.006)	(0.007)	(0.004)	(0.004)	(0.008)	(0.001)	(0.002)	(0.002)
		500	0.088	0.052	0.244	0.276	1.466	0.874	1.404	0.812
			(0.004)	(0.005)	(0.002)	(0.002)	(0.005)	(0.000)	(0.000)	(0.001)
	CLS	100	0.142	0.039	0.215	0.329	1.235	0.877	0.965	0.913
			(0.011)	(0.012)	(0.009)	(0.010)	(0.012)	(0.006)	(0.013)	(0.006)
		300	0.098	0.056	0.196	0.311	1.026	0.854	1.002	0.926
			(0.006)	(0.008)	(0.007)	(0.008)	(0.009)	(0.005)	(0.011)	(0.004)
		500	0.103	0.049	0.246	0.289	1.336	0.821	1.432	0.846
			(0.005)	(0.005)	(0.005)	(0.006)	(0.007)	(0.003)	(0.009)	(0.003)
ZI-PWE	CML	100	0.133	0.054	0.183	0.328	1.255	0.910	1.182	0.759
			(0.001)	(0.002)	(0.002)	(0.002)	(0.001)	(0.001)	(0.000)	(0.000)
		300	0.102	0.045	0.240	0.311	1.446	0.922	0.874	0.821
			(0.000)	(0.001)	(0.000)	(0.000)	(0.000)	(0.000)	(0.000)	(0.000)
		500	0.140	0.047	0.212	0.296	1.715	0.881	0.906	0.998
			(0.000)	(0.000)	(0.000)	(0.000)	(0.000)	(0.000)	(0.000)	(0.000)
	CLS	100	0.105	0.052	0.202	0.342	1.023	0.895	0.857	0.915
			(0.003)	(0.002)	(0.004)	(0.001)	(0.003)	(0.002)	(0.002)	(0.001)
		300	0.085	0.048	0.159	0.289	1.285	0.906	0.866	0.888
			(0.002)	(0.001)	(0.003)	(0.000)	(0.000)	(0.000)	(0.001)	(0.000)
		500	0.112	0.054	0.246	0.319	1.059	0.889	0.896	0.906
			(0.001)	(0.000)	(0.000)	(0.000)	(0.001)	(0.000)	(0.000)	(0.000)
ZI-PGLD	CML	100	0.114	0.055	0.236	0.309	0.968	0.889	0.986	0.866
			(0.001)	(0.000)	(0.000)	(0.000)	(0.000)	(0.000)	(0.000)	(0.001)
		300	0.101	0.048	0.156	0.256	0.846	0.857	0.879	0.973
			(0.000)	(0.000)	(0.000)	(0.000)	(0.000)	(0.000)	(0.000)	(0.000)
		500	0.107	0.053	0.206	0.274	0.875	0.926	0.889	0.877
			(0.000)	(0.000)	(0.000)	(0.000)	(0.000)	(0.000)	(0.000)	(0.000)
	CLS	100	0.137	0.045	0.243	0.315	0.926	0.966	0.958	0.789
			(0.000)	(0.000)	(0.000)	(0.000)	(0.000)	(0.000)	(0.000)	(0.000)
		300	0.110	0.055	0.169	0.266	0.889	0.876	0.974	0.956
			(0.000)	(0.000)	(0.000)	(0.000)	(0.000)	(0.000)	(0.000)	(0.000)
		500	0.126	0.059	0.195	0.321	0.916	0.959	0.865	0.909
			(0.000)	(0.000)	(0.000)	(0.000)	(0.000)	(0.000)	(0.000)	(0.000)

TABLE 3.16

Simulated estimates and corresponding standard errors in () under different Zero-Inflated Poisson-mixture innovations with Generalized Binomial thinning, assuming stationarity

Innovation	Method	T	$\rho_1=0.1$	$\rho_2=0.05$	$\rho_3=0.2$	$\rho_4=0.3$	$\beta_0=0.3$	$\beta_1=0.2$	$\eta_0=0.3$	$\eta_1=0.2$	ν	γ
ZI-Geometric	CML	100	0.104	0.052	0.160	0.280	0.336	0.213	0.318	0.224		0.878
			(0.006)	(0.007)	(0.004)	(0.005)	(0.002)	(0.004)	(0.007)	(0.008)		(0.004)
		300	0.128	0.051	0.203	0.267	0.342	0.244	0.256	0.156		0.918
			(0.005)	(0.005)	(0.003)	(0.003)	(0.001)	(0.003)	(0.003)	(0.006)		(0.003)
		500	0.110	0.053	0.234	0.322	0.276	0.229	0.302	0.187		0.875
			(0.002)	(0.003)	(0.001)	(0.002)	(0.000)	(0.001)	(0.001)	(0.004)		(0.002)
	CLS	100	0.112	0.054	0.214	0.346	0.256	0.231	0.306	0.246		0.942
			(0.012)	(0.011)	(0.009)	(0.008)	(0.009)	(0.011)	(0.009)	(0.012)		(0.009)
		300	0.098	0.047	0.189	0.312	0.276	0.219	0.289	0.213		0.889
			(0.009)	(0.009)	(0.007)	(0.006)	(0.008)	(0.009)	(0.008)	(0.009)		(0.008)
		500	0.146	0.053	0.231	0.287	0.316	0.198	0.334	0.197		0.916
			(0.006)	(0.008)	(0.005)	(0.005)	(0.005)	(0.008)	(0.005)	(0.006)		(0.007)
ZI-NB	CML	100	0.143	0.049	0.179	0.283	0.325	0.193	0.254	0.186	0.708	0.936
			(0.004)	(0.003)	(0.005)	(0.005)	(0.006)	(0.009)	(0.007)	(0.009)	(0.006)	(0.003)
		300	0.131	0.050	0.168	0.324	0.286	0.237	0.346	0.174	0.886	0.922
			(0.003)	(0.002)	(0.003)	(0.002)	(0.004)	(0.007)	(0.005)	(0.008)	(0.004)	(0.002)
		500	0.139	0.046	0.243	0.334	0.301	0.194	0.259	0.202	0.656	0.916
			(0.001)	(0.000)	(0.002)	(0.001)	(0.003)	(0.005)	(0.003)	(0.006)	(0.003)	(0.001)
	CLS	100	0.088	0.049	0.195	0.317	0.296	0.233	0.349	0.216	0.721	0.887
			(0.006)	(0.005)	(0.008)	(0.009)	(0.008)	(0.012)	(0.009)	(0.011)	(0.009)	(0.004)
		300	0.102	0.052	0.241	0.296	0.344	0.196	0.276	0.229	0.934	0.913
			(0.005)	(0.004)	(0.007)	(0.007)	(0.005)	(0.009)	(0.008)	(0.009)	(0.008)	(0.003)
		500	0.096	0.059	0.213	0.346	0.279	0.241	0.318	0.244	0.896	0.876
			(0.003)	(0.003)	(0.005)	(0.004)	(0.004)	(0.007)	(0.007)	(0.006)	(0.005)	(0.002)
ZI-CMP	CML	100	0.129	0.047	0.155	0.294	0.259	0.186	0.331	0.192	0.778	0.865
			(0.016)	(0.021)	(0.011)	(0.015)	(0.009)	(0.012)	(0.016)	(0.023)	(0.018)	(0.008)
		300	0.122	0.056	0.177	0.334	0.267	0.197	0.295	0.206	0.856	0.898
			(0.012)	(0.017)	(0.008)	(0.013)	(0.008)	(0.009)	(0.012)	(0.017)	(0.017)	(0.007)
		500	0.109	0.052	0.212	0.288	0.328	0.211	0.281	0.177	0.859	0.932
			(0.009)	(0.014)	(0.006)	(0.009)	(0.005)	(0.006)	(0.009)	(0.015)	(0.013)	(0.005)
	CLS	100	0.106	0.059	0.236	0.329	0.299	0.242	0.256	0.244	0.901	0.934
			(0.023)	(0.019)	(0.015)	(0.019)	(0.013)	(0.019)	(0.018)	(0.018)	(0.020)	(0.011)
		300	0.099	0.047	0.197	0.278	0.317	0.338	0.349	0.199	0.896	0.889
			(0.019)	(0.015)	(0.013)	(0.015)	(0.009)	(0.015)	(0.015)	(0.015)	(0.016)	(0.009)
		500	0.021	0.056	0.206	0.306	0.309	0.259	0.289	0.231	0.814	0.946
			(0.015)	(0.013)	(0.009)	(0.013)	(0.007)	(0.012)	(0.012)	(0.009)	(0.013)	(0.008)
ZI-PL	CML	100	0.122	0.048	0.183	0.247	0.319	0.176	0.326	0.231		0.915
			(0.012)	(0.015)	(0.010)	(0.016)	(0.014)	(0.009)	(0.011)	(0.008)		(0.016)
		300	0.114	0.051	0.165	0.275	0.288	0.188	0.342	0.224		0.898
			(0.009)	(0.009)	(0.007)	(0.012)	(0.009)	(0.007)	(0.007)	(0.006)		(0.014)
		500	0.130	0.050	0.159	0.296	0.324	0.194	0.296	0.168		0.923
			(0.007)	(0.007)	(0.005)	(0.009)	(0.007)	(0.005)	(0.005)	(0.005)		(0.011)
	CLS	100	0.098	0.069	0.229	0.264	0.305	0.244	0.333	0.246		0.877
			(0.019)	(0.019)	(0.009)	(0.019)	(0.021)	(0.019)	(0.015)	(0.009)		(0.025)
		300	0.111	0.059	0.196	0.302	0.289	0.219	0.290	0.226		0.916
			(0.017)	(0.018)	(0.007)	(0.018)	(0.016)	(0.013)	(0.011)	(0.007)		(0.021)
		500	0.088	0.058	0.207	0.322	0.311	0.198	0.329	0.188		0.889
			(0.018)	(0.016)	(0.005)	(0.015)	(0.014)	(0.009)	(0.009)	(0.006)		(0.017)

Innovation	Method	T	ρ_1=0.1	ρ_2=0.05	ρ_3=0.2	ρ_4=0.3	β_0=0.3	β_1=0.2	η_0=0.3	η_1=0.2	ν	γ
ZI-PT	CML	100	0.136	0.045	0.223	0.270	0.341	0.206	0.287	0.193	1.325	0.913
			(0.012)	(0.013)	(0.009)	(0.005)	(0.006)	(0.011)	(0.003)	(0.009)	(0.008)	(0.003)
		300	0.133	0.049	0.189	0.336	0.336	0.228	0.320	0.215	1.659	0.925
			(0.008)	(0.006)	(0.006)	(0.003)	(0.005)	(0.006)	(0.002)	(0.006)	(0.003)	(0.002)
		500	0.128	0.052	0.178	0.310	0.297	0.234	0.284	0.166	1.453	0.877
			(0.006)	(0.005)	(0.004)	(0.001)	(0.003)	(0.005)	(0.000)	(0.004)	(0.002)	(0.001)
	CLS	100	0.113	0.063	0.236	0.334	0.266	0.218	0.316	0.229	1.211	0.942
			(0.016)	(0.019)	(0.013)	(0.011)	(0.009)	(0.014)	(0.009)	(0.011)	(0.014)	(0.008)
		300	0.144	0.059	0.196	0.267	0.332	0.199	0.288	0.237	1.376	0.879
			(0.012)	(0.013)	(0.009)	(0.008)	(0.007)	(0.012)	(0.008)	(0.009)	(0.012)	(0.006)
		500	0.099	0.042	0.218	0.317	0.289	0.232	0.344	0.186	1.621	0.932
			(0.009)	(0.011)	(0.008)	(0.005)	(0.005)	(0.009)	(0.005)	(0.008)	(0.008)	(0.005)
ZI-PIG	CML	100	0.111	0.053	0.196	0.263	0.264	0.196	0.279	0.159		0.937
			(0.021)	(0.015)	(0.008)	(0.009)	(0.011)	(0.008)	(0.007)	(0.015)		(0.009)
		300	0.134	0.058	0.174	0.267	0.255	0.187	0.313	0.226		0.889
			(0.019)	(0.013)	(0.007)	(0.006)	(0.009)	(0.006)	(0.004)	(0.012)		(0.007)
		500	0.138	0.055	0.243	0.282	0.289	0.162	0.300	0.232		0.948
			(0.016)	(0.008)	(0.004)	(0.004)	(0.008)	(0.005)	(0.003)	(0.009)		(0.006)
	CLS	100	0.096	0.049	0.239	0.305	0.312	0.216	0.327	0.197		0.996
			(0.028)	(0.016)	(0.011)	(0.015)	(0.016)	(0.009)	(0.009)	(0.019)		(0.015)
		300	0.086	0.052	0.196	0.268	0.344	0.231	0.259	0.211		0.859
			(0.027)	(0.009)	(0.009)	(0.011)	(0.009)	(0.008)	(0.007)	(0.015)		(0.012)
		500	0.134	0.058	0.258	0.316	0.289	0.246	0.308	0.186		0.921
			(0.021)	(0.011)	(0.007)	(0.009)	(0.007)	(0.006)	(0.005)	(0.009)		(0.009)
ZI-WCG	CML	100	0.099	0.051	0.186	0.321	0.327	0.186	0.296	0.244	1.152	0.902
			(0.005)	(0.006)	(0.002)	(0.005)	(0.004)	(0.003)	(0.004)	(0.003)	(0.005)	(0.003)
		300	0.142	0.046	0.231	0.294	0.306	0.207	0.328	0.165	1.059	0.916
			(0.003)	(0.005)	(0.001)	(0.002)	(0.003)	(0.002)	(0.002)	(0.002)	(0.003)	(0.002)
		500	0.097	0.044	0.220	0.279	0.312	0.216	0.311	0.191	1.228	0.859
			(0.002)	(0.003)	(0.000)	(0.000)	(0.002)	(0.001)	(0.001)	(0.001)	(0.002)	(0.001)
	CLS	100	0.145	0.056	0.214	0.348	0.323	0.229	0.316	0.234	0.996	0.859
			(0.009)	(0.008)	(0.003)	(0.006)	(0.007)	(0.008)	(0.009)	(0.005)	(0.011)	(0.006)
		300	0.099	0.048	0.233	0.296	0.300	0.177	0.255	0.196	1.325	0.911
			(0.008)	(0.005)	(0.002)	(0.004)	(0.004)	(0.006)	(0.008)	(0.004)	(0.007)	(0.005)
		500	0.106	0.052	0.200	0.326	0.296	0.195	0.327	0.247	1.026	0.877
			(0.006)	(0.003)	(0.001)	(0.003)	(0.003)	(0.005)	(0.005)	(0.003)	(0.005)	(0.004)
ZI-PWE	CML	100	0.103	0.053	0.203	0.268	0.337	0.189	0.268	0.175	0.876	0.919
			(0.000)	(0.002)	(0.000)	(0.001)	(0.000)	(0.000)	(0.000)	(0.000)	(0.000)	(0.001)
		300	0.122	0.058	0.150	0.272	0.299	0.194	0.251	0.203	0.855	0.936
			(0.000)	(0.001)	(0.000)	(0.000)	(0.000)	(0.000)	(0.000)	(0.000)	(0.000)	(0.000)
		500	0.117	0.052	0.241	0.316	0.271	0.232	0.306	0.217	0.914	0.859
			(0.000)	(0.000)	(0.000)	(0.000)	(0.000)	(0.000)	(0.000)	(0.000)	(0.000)	(0.000)
	CLS	100	0.148	0.059	0.232	0.335	0.343	0.249	0.326	0.224	0.856	0.889
			(0.003)	(0.004)	(0.001)	(0.003)	(0.000)	(0.000)	(0.002)	(0.000)	(0.000)	(0.003)
		300	0.123	0.049	0.196	0.321	0.288	0.186	0.308	0.196	0.921	0.916
			(0.001)	(0.003)	(0.000)	(0.001)	(0.000)	(0.000)	(0.001)	(0.000)	(0.000)	(0.002)
		500	0.108	0.052	0.259	0.304	0.319	0.237	0.311	0.243	0.933	0.906
			(0.000)	(0.002)	(0.000)	(0.000)	(0.000)	(0.000)	(0.000)	(0.000)	(0.000)	(0.001)
ZI-PGLD	CML	100	0.113	0.052	0.201	0.326	0.277	0.231	0.256	0.223	0.916	0.785
			(0.000)	(0.000)	(0.000)	(0.000)	(0.000)	(0.000)	(0.000)	(0.000)	(0.000)	(0.000)
		300	0.096	0.054	0.199	0.289	0.289	0.178	0.346	0.176	0.889	0.997
			(0.000)	(0.000)	(0.000)	(0.000)	(0.000)	(0.000)	(0.000)	(0.000)	(0.000)	(0.000)
		500	0.089	0.048	0.215	0.316	0.254	0.223	0.332	0.214	0.907	0.856
			(0.000)	(0.000)	(0.000)	(0.000)	(0.000)	(0.000)	(0.000)	(0.000)	(0.000)	(0.000)
	CLS	100	0.145	0.051	0.196	0.309	0.319	0.196	0.344	0.232	0.911	0.937
			(0.000)	(0.000)	(0.000)	(0.000)	(0.000)	(0.000)	(0.000)	(0.000)	(0.000)	(0.000)
		300	0.128	0.050	0.243	0.299	0.269	0.201	0.296	0.187	0.879	0.912
			(0.000)	(0.000)	(0.000)	(0.000)	(0.000)	(0.000)	(0.000)	(0.000)	(0.000)	(0.000)
		500	0.122	0.049	0.217	0.322	0.278	0.226	0.326	0.198	0.965	0.888
			(0.000)	(0.000)	(0.000)	(0.000)	(0.000)	(0.000)	(0.000)	(0.000)	(0.000)	(0.000)

TABLE 3.17

Simulated estimates and corresponding standard errors in () under different Zero-Inflated Poisson-mixture innovations with Generalized Binomial thinning, assuming non-stationarity

PGLD innovations and their ZI versions, provided less biased estimates under both CML and CLS approaches and under stationary and non-stationary settings. However, on average, 10 to 15% of simulations fail with the CMP, PT, PL, and WCG innovations under the CML approach, while CLS does not report any non-convergent simulations. Note that even though the INAR (4) with NB, PWE, and PGLD innovations and their ZI versions under the GB thinning reaped lower standard errors, we noticed huge computational failures under the GB thinning, be it under stationary or non-stationary settings, especially at $T = 300$ and $T = 500$, where more than 300 simulations failed. The computational procedures also become very slow and time-consuming. Conversely, this computational restriction was surprisingly not observed at Section 3.2.

3.4 Concluding Remarks

This chapter provides some good insights on the workability of the different INAR processes under the Binomial and Generalized Binomial thinnings. Based on the simulated results, we retain that the INAR (4) process with the ZI-NB, ZI-PWE, ZI-PGLD, ZI-Geometric, ZI-WCG, ZI-CMP, and ZI-P innovations performed well under the Generalized Binomial and binomial thinning under both the CML and CLS approaches and under both non-stationary and stationary settings. However, as mentioned earlier, we noticed some computational failures with the CML approach, in particular with the CMP, PT, PIG, PL, and WCG innovations, but conclusively, CML provides less biased estimates with lesser standard errors than the CLS. It is proposed that the CLS approach can equally be used as an alternative to CML in cases of computational breakdowns. Working under the GB thinning proved to be computationally challenging, whilst under the binomial thinning, the simulation processes under the different Poisson-mixture models and their ZI versions in stationary and non-stationary settings are comparatively more executable. GB thinning should thus be used restrictively. Another area of observations is to involve Hurdle models in our analysis considering their flexibility to account for excess zeros in a particular series. In the next chapter, we start by applying the mentioned different ZI-Poisson mixtures innovations distributions to real life example like the SARs-COV 2 series in Mauritius under the GB, the Binomial and the less explored, NB thinning, to better assess the performance of the different INAR models and thinnings and draw up better conclusion which will then serve as benchmark for the subsequent area of research.

4

Application: The Novel Coronavirus 2019 (COVID-19) in Mauritius

4.1 Situational Assessment of COVID-19 Pandemic in Mauritius—From 2020 till 2022

Mauritius reported its first case of COVID-19 in March 2020 but has thrice been swept by the waves of the COVID-19 pandemic. With a whopping over 240,000 confirmed cases and over 1,000 death tolls, as of August 2022, the Mauritian population has now learned to manage and bear the consequences of COVID-19. Indeed, the situation was worrisome in 2020 when the COVID-19 infection and death cases were peaking, and Mauritius, due to a lack of adequate advanced health equipment and personnel, witnessed an unprecedented social and economic crisis. But hats off to the timely strict sanitary measures in terms of national lockdown, safe shopping guidelines, mandatory face covering in public places, and minimal public gathering, followed by COVID-19-related legislations like the Quarantine Bill and COVID-19 Miscellaneous Bill [Mamode Khan et al., 2020b], which surely slowed the propagation rate of SARS-CoV-2. Most importantly, with the emergence of several highly effective vaccines, and via its strong bilateral agreements, policy makers in Mauritius ensured that herb immunity is achieved by all the Mauritian's population especially old-aged persons, people with comorbidities, and personnel working in the education, retail and other economic sectors. In 2022, vaccination campaigns are still on, given the need to get a second 'booster' dose and re-enforce the immune system.

Throughout the COVID-19 era in Mauritius, different sanitary measures, like the never-implemented-before novel localised mobility restrictions in regions ("red-zoned" areas) under the Temporary Restrictions of Movement Order and new COVID-19 focused legislations to enforce vaccination, were implemented and all these along with some unmeasurable effects like environmental factors, the responsiveness or adherence of the Mauritian population amongst others, contributed in containing the virus rapidly and strengthening the health system.

In terms of evidenced based policies, Mauritius is indeed well positioned but on the other hand, the uncommon yet interesting patterns in the COVID-19 series, see Figure 1.1, raise some concerns in the research community especially in the midst of statistics and data analytics. As already discussed, the COVID-19 new cases series in Mauritius has some distinctive features like a purely unique serial trend with excess of zeros and some oscillations, leading to over-dispersion, while the corresponding

DOI: 10.1201/9781003677451-4

COVID-19-related death series describes a preponderance of excess zeros. These series thus imply that the simple INAR (1) model with Poisson or extra-Poisson innovations is surely insufficient in this context [McKenzie, 1986, 1988, Mamode Khan et al., 2021] and ignoring the excess of zeros will lead to biasedness in the estimated parameters and standard errors [Zuur et al., 2009]. To remedy, it is proposed to construct a novel high-ordered INAR process with the ZI innovation distributions, reported under Chapter 3 (ZI-PWE, ZI-PGLD, ZI-Geometric, ZI-NB, ZI-WCG, ZI-CMP and ZI-P) under the Generalized Binomial and the Negative Binomial thinning. More on the formulation of ZI models has already been described at Chapter 3.

This novel model bridges three important research gaps. Firstly, as seen in the literature, the ZI models have extensively been used in regression contexts only (See Lambert [1992], Perumean-Chaney et al. [2012] and the references therein) while its applications in counting time series modeling is quite restricted to first order only [Jazi et al., 2012, Bareto Souza, 2015, Zhu, 2012a, Goncalves et al., 2016, Moller et al., 2017, Bakouch et al., 2017, Conceicao et al., 2017, Bourgignon et al., 2018, Bourguignon, 2018, Moller et al., 2017, 2018]. Secondly, the proposed time series model allows for covariate specification, which in the context of the COVID-19 analysis, is primordial. In fact, it is important to identify the significant factors contributing to the propagation of SARS-CoV-2 in the local community, while also detecting the expected impact of the covariates on the COVID-19 infection in the local community. Thereon, such information will extremely be useful to the local concerned authorities and for forecasting purposes. As regard to the factors, in the first wave of COVID-19 in Mauritius, we concluded that public health measures, strong political engagement, stricter legislations, population behavior to established sanitary norms [Musango et al., 2021], sensitization campaigns and the institution of quarantine centers successfully helped in curbing the spread of the virus [Mamode Khan et al., 2020b]. Now, with aim to assess the second and third waves of the COVID-19, new covariates like the reproduction rate (ReR), the COVID-19 Risk due to weather conditions (CRW), the major event of vaccination and the COVID-19 Stringency Index have been considered in the Mauritian context. To note that, in the European and Asian regions, ReR [Linka et al., 2020, Chudik et al., 2021, Tyagi et al., 2020], CRW [Guo et al., 2020, Xu et al., 2020] and vaccinations [Shah et al., 2021] have largely demonstrated their association with COVID-19 transmission but for Mauritius, the COVID-19 Stringency index was the most significant factor in curbing the virus [Mamode Khan et al., 2021] during the first wave. More description on the mentioned covariates is provided in Section 4.2 of this Chapter 4. An accurate forecasting with acceptable RMSE is also targeted because most restorative and preparedness policy decisions, be it in terms of adequate vaccines, financial requirement and re-opening of frontier, will be based on the new COVID-19 cases' projections.

Finally, the use of the GB and NB thinning under high-ordered INAR processes required more exploration. Indeed, Nastic and Bakouch [2012] worked on the NB thinning firstly proposed by Ristic et al. [2009, 2012] and introduced a combined geometric INAR(p) model based on negative binomial thinning but, as of now, its applicability under non-stationary setting for INAR(p) with zero-inflated

Poisson-mixtures models is yet to be explored. This work thus intend to add value to the literature.

4.2 Application Results and Discussion

This section focuses on fitting the different INAR models to the SARS-CoV-2 series in Mauritius. To account for the non-stationarity, we assume the following time-dependent covariates:

- The COVID-19 Stringency Index (SI): This variable has been calculated from nine metrics, namely school closures; workplace closures; cancellation of public events; restrictions on public gatherings; closures of public transport; stay-at-home requirements; public information campaigns; restrictions on internal movements; and international travel controls and is available on a daily basis at [1]. This index gives an indication of the strictness of government policies to the COVID-19 pandemic. The score is between 0 to 100 where an index nearing 100 indicates strict response otherwise less strict response [Ritchie et al., 2021]. In this study, the logarithm of the nominal value of the SI (log(SI)) was used for analysis purposes.

- The event of vaccination (Vaccine): This is an important time-varying variable because as at date, the vaccine roll-outs in Mauritius is rising given the authorities's aim to achieve herd immunity. More than 1 million vaccine doses have already been obtained through bilateral agreements with India and China even though there is an intense competition between countries for the purchase the COVID-19 vaccine. For this study, the event of vaccination was categorised into two possibilities (binary) where 1 indicates that the event for vaccination is being done and 0 (ref) that the event of vaccination is not being done. Data on vaccination was obtained from [2] [Jeewa, 2021] and from the official COVID-19 platform in Mauritius, falling under the aegis of the Ministry of Health and Wellness [3].

- The reproduction rate (ReR): This covariate refers to the degree of propagation of SARS-CoV-2 from one person to another. The data on ReR was obtained from [4] and the logarithm of the nominal value of ReR (log(ReR)) was considered. To note that this reproduction rate relates to the degree of transmissibility of the "original" SARS-CoV-2 and was used as a proxy to understand the severity of the coronavirus, considering that nowadays, constant mutation of the SARS-CoV-2 has been observed. In terms of proactive health related measures, this rate can be highly indicative. The logarithm of the nominal value of ReR (log(ReR)) was considered.

---

[1] https://ourworldindata.org/covid-stringency-index
[2] https://ourworldindata.org/covid-vaccinations
[3] https://covid19.mu/
[4] https://ourworldindata.org/covid-cases

• The Relative COVID-19 Risk due to Weather and Air Pollution (CRW): This variable represents some environmental factors and explains their impact on COVID-19 transmission. Weather factors like average and diurnal temperature, ultraviolet (UV) index, humidity, pressure, precipitation and air pollutants (SO2 and Ozone) were considered while computing this index. In this book, the CRW was categorised into 0 and 1 (binary) where 0 (ref) is when an index is below 1 referring to relatively lower impact of weather factors on spread of COVID-19 and 1 is when an index is above 1, indicating otherwise. Data has been extracted from and imputation based on observations were done for missing values.

The estimates of the above variables are shown in Tables 4.1 to 4.3:

The results in Tables 4.1 to 4.4, were obtained assuming the training dataset from 18 March 2020 to 25 April 2021. It can be deduced that the Akaike Information Criteria (AIC) depends hugely on the type of thinning operator being used. The application of different zero-inflated Poisson-mixtures innovations under the Negative Binomial and Generalized Binomial thinning to the COVID-19 series in Mauritius proved to be computationally complex however comparatively, non-stationary INAR (7) with ZI-NB, ZI-PGLD, ZI-PWE, ZI-Geometric and ZI-CMP under the Generalized Binomial thinnings provided the lower AIC. These ZI- Poisson mixtures models under the binomial thinning were executed with no computational failures but does not fit well to the COVID-19 new infection series of Mauritius since its AIC is relatively higher. Based on the estimates obtained from the ZI-NB model, the variables 'ReR', 'SI', and 'Vaccine' were highly significant in reducing the number of infection in Mauritius, as compared to 'CRW'.

The ReR is directly associated with the number of new active cases Billah et al. [2020]. This is because by observing the evolution of the series, it can be deduced that in October 2020 when there were an increase in international mobility Linka et al. [2020] following opening of frontier and in March 2021 when the second wave has resurfaced, an exponential increase in the number of active COVID-19 cases was reported. At this point, a worrisome 'ReR' of above 1 was being reported, indicating high risk of getting infected. Fortunately, based on these trends in 'ReR' and new COVID-19 active cases, the authorities triggered timely health related measures like vaccination campaigns in Mauritius and consequently, the policies proved its effectiveness in April 2021, with a reduction in the number of new active COVID-19 cases, and in the 'ReR'. At this point, 'ReR' was below 0.5.

The event of vaccination indeed is playing a vital role in curbing the number in infection. Based on the reversed estimates of 'Vaccines', it can be deduced that as the vaccination campaigns take place, this is reflected positively in the share of Mauritian population which has already received at least one dose of the vaccines and likewise, the risk of getting infected is expected to decrease considerably. It has largely been proven that the COVID-19 vaccines reduces the overall attack rate by rendering the human immunity system more resilient. The chance for symptomatic and asymptomatic infections Haas et al. [2021], BBC [2021], Dyer [2021], Iacobucci [2021], Lorenz [2021], Murphy [2021] and the severity of the symptoms Juno and Wheatley

Innovation	Thinning	Parameter	ρ_1	ρ_2	ρ_3	ρ_4	ρ_5	ρ_6	ρ_7	Intercept	ReR	SI	Vaccine	CRW	η_0	η_1	η_2	η_3	η_4	γ	AIC	LL
ZI-P	Binomial	Estimates	0.204	0.107	0.001	0.056	0.042	0.041	0.031	0.397	0.092	-0.404	-0.648	0.253	0.168	0.038	0.303	0.056	0.517		9081.5	9047.5
		Std Error	(0.000)	(0.000)	(0.001)	(0.002)	(0.005)	(0.004)	(0.010)	(0.001)	(0.000)	(0.000)	(0.000)	(0.001)	(0.001)	(0.000)	(0.009)	(0.009)	(0.001)			
		p-value	0.000	0.000	0.000	0.000	0.000	0.000	0.001	0.000	0.000	0.000	0.000	0.000	0.000	0.000	0.000	0.000	0.000			
	GB	Estimates	0.206	0.107	0.104	0.053	0.043	0.042	0.033	0.287	0.331	-0.372	-0.641	0.214	0.262	0.174	0.188	0.145	0.154	1.083	8759.0	8725.0
		Std Error	(0.000)	(0.000)	(0.002)	(0.001)	(0.009)	(0.012)	(0.008)	(0.000)	(0.006)	(0.000)	(0.000)	(0.110)	(0.008)	(0.020)	(0.021)	(0.020)	(0.495)	(0.000)		
		p-value	0.000	0.000	0.000	0.000	0.000	0.001	0.000	0.000	0.000	0.000	0.000	0.051	0.000	0.000	0.000	0.000	0.755	0.000		
	NB	Estimates	0.222	0.096	0.110	0.052	0.043	0.041	0.027	0.483	0.033	-0.458	-0.364	0.301	0.303	0.235	0.026	0.133	0.081		8953.1	8919.1
		Std Error	(0.000)	(0.010)	(0.006)	(0.021)	(0.010)	(0.011)	(0.021)	(0.001)	(0.000)	(0.000)	(0.001)	(0.211)	(0.018)	(0.024)	(0.032)	(0.012)	(0.483)			
		p-value	0.000	0.000	0.000	0.012	0.000	0.000	0.184	0.000	0.000	0.000	0.000	0.153	0.000	0.000	0.412	0.000	0.866			
ZI-Geometric	Binomial	Estimates	0.186	0.094	0.089	0.068	0.030	0.036	0.031	0.565	0.163	0.504	-0.748	0.122	-0.014	0.503	-0.193	-0.100	0.324		3607.8	3573.8
		Std Error	(0.019)	(0.006)	(0.035)	(0.081)	(0.060)	(0.016)	(0.007)	(0.008)	(0.001)	(0.011)	(0.011)	(0.039)	(0.006)	(0.009)	(0.016)	(0.018)	(0.013)			
		p-value	0.000	0.000	0.011	0.402	0.620	0.026	0.000	0.000	0.000	0.000	0.000	0.000	0.033	0.000	0.000	0.000	0.000			
	GB	Estimates	0.202	0.095	0.113	0.051	0.038	0.051	0.045	0.278	0.164	-0.459	-1.043	0.167	-0.086	0.184	0.075	0.163	0.121	0.842	3269.8	3235.8
		Std Error	(0.034)	(0.008)	(0.008)	(0.012)	(0.004)	(0.008)	(0.004)	(0.000)	(0.000)	(0.000)	(0.000)	(0.301)	(0.003)	(0.000)	(0.000)	(0.003)	(0.901)	(0.014)		
		p-value	0.000	0.000	0.000	0.000	0.000	0.000	0.000	0.000	0.000	0.000	0.000	0.578	0.000	0.000	0.000	0.000	0.893	0.000		
	NB	Estimates	0.194	0.093	0.063	0.044	0.035	0.025	0.069	0.836	0.184	-0.407	-1.094	0.581	-0.143	0.534	0.485	0.180	0.556		4014.1	3980.1
		Std Error	(0.000)	(0.000)	(0.000)	(0.000)	(0.000)	(0.000)	(0.000)	(0.000)	(0.000)	(0.000)	(0.001)	(0.411)	(0.000)	(0.000)	(0.000)	(0.000)	(0.801)			
		p-value	0.000	0.000	0.000	0.000	0.000	0.000	0.000	0.000	0.000	0.000	0.000	0.157	0.000	0.000	0.000	0.000	0.488			
ZI-NB	Binomial	Estimates	0.188	0.114	0.111	0.056	0.042	0.041	0.041	0.095	0.063	-0.098	-0.107	0.034	0.028	0.17	-0.369	0.149	0.232		3548.8	3514.8
		Std Error	(0.000)	(0.000)	(0.000)	(0.000)	(0.000)	(0.000)	(0.000)	(0.001)	(0.000)	(0.000)	(0.001)	(0.128)	(0.001)	(0.000)	(0.000)	(0.001)	(0.125)			
		p-value	0.000	0.000	0.000	0.000	0.000	0.000	0.000	0.000	0.000	0.000	0.000	0.788	0.000	0.000	0.000	0.000	0.063			
	GB	Estimates	0.319	0.144	0.171	0.140	0.021	0.113	0.055	0.330	0.452	-0.308	-0.467	0.266	-0.846	0.625	0.505	0.100	0.180	0.852	2493.1	2459.1
		Std Error	(0.003)	(0.015)	(0.025)	(0.046)	(0.006)	(0.001)	(0.001)	(0.001)	(0.040)	(0.038)	(0.000)	(0.194)	(1.656)	(0.149)	(0.357)	(0.009)	(0.694)	(0.023)		
		p-value	0.000	0.000	0.000	0.003	0.000	0.000	0.000	0.000	0.000	0.000	0.000	0.170	0.609	0.000	0.157	0.000	0.795	0.000		
	NB	Estimates	0.206	0.089	0.077	0.054	0.044	0.046	0.055	0.262	0.158	-0.308	-0.023	0.253	-0.189	0.390	0.401	0.752	0.122		4085.4	4051.4
		Std Error	(0.001)	(0.002)	(0.002)	(0.002)	(0.002)	(0.002)	(0.001)	(0.619)	(0.001)	(0.002)	(0.006)	(0.149)	(1.026)	(0.001)	(0.001)	(0.026)	(0.276)			
		p-value	0.000	0.000	0.000	0.000	0.000	0.000	0.000	0.672	0.000	0.000	0.000	0.089	0.854	0.000	0.000	0.000	0.660			

TABLE 4.1

Results under Binomial, GB, and NB thinnings for ZI-P, ZI-Geometric, and ZI-NB models

Innovation	Thinning	Parameter	ρ_1	ρ_2	ρ_3	ρ_4	ρ_5	ρ_6	ρ_7	Intercept	ReR	SI	Vaccine	CRW	η_0	η_1	η_2	η_3	η_4	γ	AIC	LL
ZI-PT	Binomial	Estimates	0.017	0.013	0.013	0.055	0.039	0.039	0.039	0.040	0.019	-0.062	-0.022	0.075	-0.070	0.075	-0.024	-0.122	0.011		8953.1	8919.1
		Std Error	(0.004)	(0.007)	(0.017)	(0.000)	(0.001)	(0.006)	(0.007)	(0.065)	(0.000)	(0.001)	(0.007)	(0.056)	(0.285)	(0.004)	(0.027)	(0.285)	(0.715)			
		p-value	0.000	0.000	0.000	0.000	0.000	0.000	0.000	0.000	0.000	0.000	0.000	0.000	0.000	0.000	0.000	0.000	0.000			
	GB	Estimates	0.052	0.102	0.041	0.116	0.240	0.365	0.127	0.045	0.103	-0.265	-0.310	0.011	0.052	0.106	0.226	0.042	0.055	1.023	8623.4	8589.4
		Std Error	(0.001)	(0.001)	(0.001)	(0.002)	(0.004)	(0.001)	(0.000)	(0.001)	(0.001)	(0.000)	(0.003)	(0.126)	(0.000)	(0.000)	(0.000)	(0.000)	(0.001)	(0.002)		
		p-value	0.000	0.000	0.000	0.000	0.000	0.000	0.000	0.000	0.000	0.000	0.000	0.930	0.000	0.000	0.000	0.000	0.000	0.000		
	NB	Estimates	0.022	0.056	0.149	0.000	0.025	0.011	0.039	0.112	0.125	-0.130	-0.161	0.215	0.045	0.056	0.036	0.011	0.102	0.000	9012.2	8978.2
		Std Error	(0.003)	(0.001)	(0.000)	(0.002)	(0.000)	(0.001)	(0.000)	(0.000)	(0.000)	(0.001)	(0.001)	(0.421)	(0.000)	(0.000)	(0.001)	(0.000)	(0.002)			
		p-value	0.000	0.000	0.000	0.000	0.000	0.000	0.000	0.000	0.000	0.000	0.000	0.610	0.000	0.000	0.000	0.000	0.000			
ZI-CMP	Binomial	Estimates	0.040	0.009	0.044	0.027	0.047	0.023	0.062	0.705	0.462	-0.399	-0.196	0.162	-0.362	2.645	-1.536	2.045	2.216		7688.6	7654.6
		Std Error	(0.001)	(0.000)	(0.376)	(0.565)	(0.001)	(0.565)	(0.376)	(0.144)	(0.000)	(0.001)	(0.044)	(0.267)	(5.735)	(1.768)	(0.000)	(5.793)	(4.982)			
		p-value	0.000	0.000	0.907	0.962	0.000	0.967	0.869	0.000	0.000	0.000	0.000	0.545	0.950	0.135	0.000	0.724	0.656			
	GB	Estimates	0.181	0.090	0.037	0.073	0.280	0.026	0.284	3.055	0.054	-0.434	-0.342	0.165	-1.391	0.453	1.970	0.672	0.563	0.891	7640.2	7606.2
		Std Error	(0.005)	(0.008)	(0.003)	(0.004)	(0.041)	(0.004)	(0.001)	(0.004)	(0.004)	(0.001)	(0.006)	(0.373)	(0.004)	(0.005)	(0.012)	(0.001)	(0.793)	(0.004)		
		p-value	0.000	0.000	0.000	0.000	0.000	0.000	0.000	0.000	0.000	0.000	0.000	0.658	0.000	0.000	0.000	0.000	0.478	0.000		
	NB	Estimates	0.030	0.051	0.038	0.041	0.022	0.031	0.023	1.480	0.401	-0.511	-0.708	0.264	0.387	1.796	0.578	0.270	0.202	0.000	7704.5	7670.5
		Std Error	(0.012)	(0.005)	(0.005)	(0.013)	(0.002)	(0.017)	(0.008)	(0.001)	(0.001)	(0.001)	(0.001)	(0.303)	(0.003)	(0.002)	(0.001)	(0.001)	(0.650)			
		p-value	0.000	0.000	0.000	0.000	0.000	0.000	0.505	0.000	0.000	0.000	0.000	0.383	0.000	0.000	0.000	0.000	0.756			
ZI-WCG	Binomial	Estimates	0.190	0.099	0.099	0.050	0.045	0.040	0.029	0.199	0.157	-0.265	-0.380	0.171	-0.069	0.240	0.197	0.344	0.074		8344.6	8310.6
		Std Error	(0.001)	(0.001)	(0.044)	(0.000)	(0.000)	(0.001)	(0.044)	(0.015)	(0.000)	(0.000)	(0.006)	(0.067)	(0.067)	(0.000)	(0.001)	(0.067)	(0.260)			
		p-value	0.000	0.000	0.024	0.000	0.000	0.000	0.000	0.000	0.000	0.000	0.000	0.011	0.309	0.000	0.000	0.000	0.775			
	GB	Estimates	0.208	0.102	0.102	0.051	0.042	0.040	0.030	0.161	0.133	-0.237	-0.216	0.245	0.093	0.223	0.286	0.300	0.950	1.009	8113.8	8079.8
		Std Error	(0.001)	(0.000)	(0.000)	(0.000)	(0.000)	(0.000)	(0.001)	(0.006)	(0.000)	(0.000)	(0.006)	(0.256)	(0.021)	(0.001)	(0.000)	(0.023)	(0.943)	(0.000)		
		p-value	0.000	0.000	0.000	0.000	0.000	0.000	0.000	0.000	0.000	0.000	0.000	0.340	0.000	0.000	0.000	0.000	0.314	0.000		
	NB	Estimates	0.200	0.098	0.101	0.051	0.041	0.040	0.030	0.119	0.225	-0.219	-0.312	0.106	0.107	0.217	0.203	0.310	0.818	0.000	8201.1	8167.1
		Std Error	(0.085)	(0.034)	(0.041)	(0.072)	(0.002)	(0.078)	(0.068)	(0.003)	(0.000)	(0.000)	(0.004)	(0.110)	(0.002)	(0.001)	(0.054)	(0.012)	(0.785)			
		p-value	0.000	0.000	0.000	0.000	0.000	0.000	0.000	0.000	0.000	0.000	0.000	0.337	0.000	0.000	0.000	0.000	0.297			

TABLE 4.2

Results under Binomial, GB, and NB thinnings for ZI-PT, ZI-CMP, and ZI-WCG models

Innovation	Parameter	Thinning	ρ_1	ρ_2	ρ_3	ρ_4	ρ_5	ρ_6	ρ_7	Intercept	ReR	SI	Vaccine	CRW	η_0	η_1	η_2	η_3	η_4	γ	AIC	LL
ZI-PWE	Estimates	Binomial	0.533	0.805	0.708	0.525	0.607	0.710	0.307	0.381	0.941	-0.374	-0.710	0.457	0.591	0.447	0.222	0.098	0.830		3071.9	3038.0
	Std Error		(0.001)	(0.001)	(0.002)	(0.003)	(0.002)	(0.001)	(0.003)	(0.003)	(0.004)	(0.000)	(0.002)	(0.904)	(0.014)	(0.005)	(0.003)	(0.014)	(0.514)			
	p-value		0.000	0.000	0.000	0.000	0.000	0.000	0.000	0.000	0.000	0.000	0.000	0.613	0.000	0.000	0.000	0.000	0.106			
	Estimates	GB	0.197	0.110	0.106	0.053	0.044	0.047	0.140	0.199	0.422	-0.101	-0.521	0.116	0.302	0.143	0.357	0.014	0.142	1.520	2747.5	2713.5
	Std Error		(0.003)	(0.003)	(0.005)	(0.001)	(0.005)	(0.504)	(0.015)	(0.066)	(0.000)	(0.001)	(0.007)	(0.251)	(0.019)	(0.003)	(0.002)	(0.002)	(0.326)	(0.018)		
	p-value		0.000	0.000	0.000	0.000	0.000	0.000	0.000	0.003	0.000	0.000	0.000	0.643	0.000	0.000	0.000	0.000	0.663	0.000		
	Estimates	NB	0.202	0.025	0.003	0.077	0.057	0.027	0.049	-0.543	-0.534	-0.278	-1.941	1.044	-2.706	-0.029	1.056	0.312	0.894		5647.8	5613.8
	Std Error		(0.009)	(0.000)	(0.000)	(0.000)	(0.001)	(0.002)	(0.002)	(0.008)	(0.000)	(0.000)	(0.008)	(0.881)	(0.000)	(0.023)	(0.000)	(0.074)	(0.557)			
	p-value		0.000	0.000	0.000	0.000	0.000	0.000	0.000	0.000	0.000	0.000	0.000	0.236	0.000	0.210	0.000	0.000	0.108			
ZI-PGLD	Estimates	Binomial	0.489	0.189	0.088	0.116	0.018	0.111	0.102	0.123	0.132	-0.174	-0.239	1.098	0.202	0.145	0.350	0.478	0.996		3578.4	3544.4
	Std Error		(0.002)	(0.001)	(0.002)	(0.001)	(0.001)	(0.002)	(0.001)	(0.002)	(0.001)	(0.001)	(0.002)	(0.985)	(0.001)	(0.002)	(0.002)	(0.003)	(0.878)			
	p-value		0.000	0.000	0.000	0.000	0.000	0.000	0.000	0.000	0.000	0.000	0.000	0.265	0.000	0.000	0.000	0.000	0.257			
	Estimates	GB	0.521	0.208	0.265	0.089	0.135	0.415	0.261	0.206	0.274	-0.483	-0.365	0.897	0.019	0.089	0.468	0.562	1.023	1.215	3065.1	3031.1
	Std Error		(0.001)	(0.000)	(0.000)	(0.000)	(0.001)	(0.001)	(0.002)	(0.001)	(0.000)	(0.002)	(0.001)	(0.652)	(0.000)	(0.000)	(0.003)	(0.002)	(0.789)	(0.003)		
	p-value		0.000	0.000	0.000	0.000	0.000	0.000	0.000	0.000	0.000	0.000	0.000	0.169	0.000	0.000	0.000	0.000	0.195	0.000		
	Estimates	NB	0.119	0.234	0.063	0.158	0.208	0.358	0.166	0.462	0.356	-0.115	-0.223	1.236	0.122	0.325	0.149	0.312	0.963		4982.5	4948.5
	Std Error		(0.002)	(0.001)	(0.000)	(0.001)	(0.002)	(0.000)	(0.001)	(0.002)	(0.001)	(0.000)	(0.001)	(0.996)	(0.001)	(0.001)	(0.001)	(0.001)	(0.966)			
	p-value		0.000	0.000	0.000	0.000	0.000	0.000	0.000	0.000	0.000	0.000	0.000	0.215	0.000	0.000	0.000	0.000	0.319			

TABLE 4.3
Results under Binomial, GB, and NB thinnings for ZI-PWE and ZI-PGLD models

Innovation	Other parameters	Thinning	Results	Innovation	Other parameters	Thinning	Results
ZI-NB	$\hat{v}$	Binomial	0.903	ZI-WCG	$\hat{\theta}$	Binomial	0.223
			(0.000)				(0.000)
			0.000				0.000
		GB	0.735			GB	0.557
			(0.007)				(0.000)
			0.000				0.000
		NB	0.844			NB	1.209
			(0.001)				(0.000)
			0.000				0.000
ZI-PT	$\hat{a}$	Binomial	1.582	ZI-PWE	$\hat{v}$	Binomial	0.930
			(0.001)				(0.000)
			0.000				0.000
		GB	1.103			GB	0.902
			(0.000)				(0.001)
			0.000				0.010
		NB	1.327			NB	0.862
			(0.002)				(0.000)
			0.000				0.000
ZI-PT	$\hat{\sigma}^2$	Binomial	0.076	ZI-PLGD	$\hat{v}$	Binomial	0.890
			(0.004)				(0.001)
			0.000				0.000
		GB	0.085			GB	0.906
			(0.002)				(0.002)
			0.000				0.000
		NB	0.072			NB	0.877
			(0.004)				(0.001)
			0.000				0.000
ZI-CMP	$\hat{v}$	Binomial	1.058				
			(0.001)				
			0.000				
		GB	0.817				
			(0.001)				
			0.000				
		NB	0.916				
			(0.001)				
			0.000				

TABLE 4.4
Results for other parameters under Binomial, GB, and NB thinnings for all ZI innovation distributions

[2021], Centers for Disease Control and Prevention [2021] are considerably reduced, thus entailing an adverse effect on the mortality rate related to COVID-19. More elaborated comments are provided below. Timely imposition of new immediate sanitary measures during the peak COVID-19 phases also play an important role in curbing

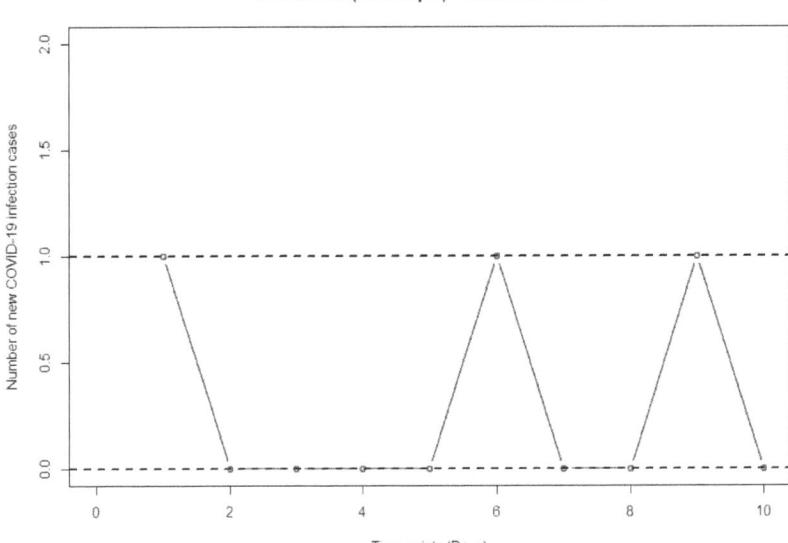

Forecasted (outsample) values with 95% CI

FIGURE 4.1
Forecasted values (out-sample) with 95% Confidence Interval

the spread of the virus. In fact, the quicker and earlier the sanitary measures are imposed, the more rapidly is the SARS-CoV-2 contained in the local community. Conversely, unlike other European regions, Mauritius reported its highest cases of COVID-19 in both warmer and colder regions, and in both weather conditions—summer and Winter, so 'CRW' was proven to be insignificant in curbing the number of active COVID-19 cases. In fact, given the constant mutation of the SARS-CoV-2 in different regions and Mauritius having a comparatively restricted regional disparity, possibly a larger dataset on 'CRW' will allow better exploration of its association with the number of infection Dobricic et al. [2020], Bukhari et al. [2020]. Table 4.4 confirms that the estimates of the over-dispersion parameters in the ZI models are significant.

Finally, the regression estimates for ZI-NB, ZI-PGLD and ZI-PWE under GB thinning, were used to conduct short term outsample forecasting of the number of new COVID-19 infection cases in Mauritius, from 26 April 2021 to 05 May 2021.

The ZI-NB model had the relatively lower Root Mean Square Errors (RMSEs) of 3.61, compared to the ZI-PGLD and ZI-PWE, which respectively had RMSEs of 4.12 and 4.36. It can also be seen that the 95% confidence interval lies between 0 and 1, which means that during the next 10 days, that is, from 26 April 2021 till 05 May 2021, there was a 95% chance that the new COVID-19 infection case will lie between 0 and 1. In Figure 4.1, we demonstrate the 95% confidence interval plot: In this book, it was interestingly found that ZI-NB outperformed the ZI-PWE and

ZI-PGLD; however, the two latter-mentioned models' results were satisfactorily adequate, which again motivates us to explore them further under possibly different types of data sets or data applications. Also, through the forecasting exercise, it was found that due to some unmeasurable and unpredictable latent effects, the forecasted number of locally acquired new COVID-19 cases may not be easily estimated, especially in the long-run. In fact, in the rise of a sudden shock or spike, the forecasted values are naturally disrupted, since the predictor functions in the innovation distribution may not include a new physical or latent effect. Such a situation may be circumvented by updating the list of covariates on a daily basis and also by allowing the forecasts on a changepoint basis. Simultaneously, it is important to check the Variance Inflating factor (VIF) of the different regressors to avoid any multi-collinearity. Likewise, in the above analysis, the factor time was omitted due to the high VIF. We also note that in the high-ordered INAR process, the specification of latent effects may not be easily handled due to integrating the random effects (refer to Mamode Khan et al. [2020b]).

4.3   Concluding Remarks

Over-dispersion is one of the many features of the COVID-19 series in Mauritius, and in the previous Chapter 3, using various ZI-Poisson mixtures innovation distribution models, we have worked on capturing the best-suited model, which is the ZI-NB, followed by ZI-PGLD and ZI-PWE under GB thinning, and have been able to conclude satisfactorily reliable results. In the next Chapter 5, we thus study another important feature which has been graphically represented in Figures 1.2 and 1.5—the periodicity. This feature in the COVID-19 series has primarily been studied by Doukhan et al. [2021] for the France and Germany COVID-19 series however, there is yet any in-depth study which has accounted for the harmonic effects or periodicity under high-ordered INAR processes with different ZI-Poisson-mixtures innovation distributions under non-stationary setting. For smooth execution of this novel model, we shall focus on the binomial thinning which proved to computationally efficient as shown in Chapter 3 and 4. Also, catering for periodicity under ZI-Poisson mixtures innovation distributions involves estimation of additional parameters; thus, as a means to adhere to the rule of parsimony, the use of binomial thinning for periodic ZI-Poisson mixture models in the subsequent chapters seemed more appropriate.

5

High-Ordered Integer-Valued Time Series Models with Harmonic Features

Examples of integer-valued series with periodic structure can include the monthly counts of claims of short-term disability benefits Freeland [1998], Fokianos [2012], the day-time and night-time road accidents Pedeli and Karlis [2011], Brijs et al. [2008], Karlis et al. [2008], the number of infected cases due to the outbreak of a virus Ferland et al. [2006], Chan et al. [2021] and, the monthly number of short-term unemployed people Monteiro et al. [2015]. In this book, in Chapter 1, it can clearly be observed that in Figure 1.3, that is in the COVID-19 new infection and death series in South Africa, there are huge fluctuations with a number of ups and downs causing oscillations and with repeated trend causing periodicity. On the other hand, in Figure 1.1, the COVID-19 new infection and death series in Mauritius exhibit huge variability over the mean, thus causing severe over-dispersion due to the preponderance of zeros in the early stage of detection. The question of interest is therefore— *How to model such series in the presence of the phenomena of periodicity, over-dispersion, presence of excess zeros, covariate specifications and others?* Up to this extent, there exists integer-valued time series models of auto-regressive nature that handles the modeling of counting time series. However, these auto-regressive models are restricted as they cannot accommodate for the above features in an unified framework. Thus, it is proposed to bring forward a flexible high-ordered integer-valued auto-regressive time series process that has the ability to simultaneously consider periodicity via harmonic specification and all sources of over-dispersion through the proper specification of the components of the time series model, while considering the simple binomial thinning with constant coefficient.

5.1 Periodic Integer-Valued Auto-regressive Models with Order p

Bentarzi and Aries [2020] did introduce the periodic integer-valued auto-regressive moving average model (PINARMA), which has the same INARMA structure as in McKenzie [1986]; that is, the current observation is related to the previous-lagged observations through the binomial thinning operator of Steutel and Van Harn [1979]. Basically, the PINARMA can suitably account for the non-stationarity with respect

DOI: 10.1201/9781003677451-5

to the moments of the counting series. This is explained in detail in Bentarzi and Aries [2020], Filho et al. [2021], Bentarzi and Bentarzi [2017], Aries and Mamode Khan. In our context, the model we consider is based on the INAR part of the PIN-ARMA model involving the use of the operator of Steutel and Van Harn [1979] and hence can generalize the stationary integer autoregressive process (*INAR*) to periodically correlated counting series, where a periodically correlated Integer-Valued process $\{Y_t, \, t \in \mathbb{Z}\}$ in the sense of Gladyshev [1963], with period S (where $S \geq 2$), is expressed as

$$Y_t = \rho_{1,t} * Y_{t-1} + \cdots + \rho_{p,t} * Y_{t-p} + R_t, \, t \in \mathbb{Z} \tag{5.1}$$

where $\{R_t, \, t \in \mathbb{Z}\}$ is a sequence of uncorrelated non-negative integer-valued random variables, with a periodic mean $\lambda_{R_t,t}$ and a finite periodic variance $\sigma^2_{R_t,t}$, and where the parameters $\rho_{i,t}$ lie in the interval $(0,1)$ and $\theta_{j,t}$ are defined for $i = 1, \ldots, p$ and $j = 1, \ldots, q$. The mean and variance of R_t are periodic in t, with period S ($S \geq 2$), such that, $\rho_{i,t+rS} = \rho_{i,t}$, $\lambda_{R_t,t+rS} = \lambda_{R_t,t}$ and $\sigma^2_{R_t,t+rS} = \sigma^2_{R_t,t}$, $\forall t, r \in \mathbb{Z}$. The counting sequences of independent non-negative integer-valued random variables $\{Y_{k,t}, \, l \in \mathbb{N}, \, t \in \mathbb{Z}\}$, where $P\left(Y_{k,t} = 1\right) = 1 - P\left(Y_{k,t} = 0\right) = \rho_{i,t} \in [0,1]$ is given by

$$\rho_{i,t} * Y_{t-i} = \begin{cases} \sum_{i=1}^{Y_{t-i}} b(\rho_{i,t}), & \text{if } Y_{t-i} > 0, \\ 0, & \text{if } Y_{t-i} = 0, \end{cases}$$

where $b(.)$ is a Bernoulli random variable with probability of success $\rho_{i,t}$ ($0 < \rho_{i,t} \leq 1$). The same definition of $*$ applies to Equation 2.1. From Du and Li [1991], the Bernoulli sequences in the paired terms, that is $(\rho_{i,t} * Y_{t-i}, \rho_{i',t} * Y_{t-i'})$ are treated independent.

Until now, periodic INAR models have been mostly used in the literature, subject to some drawbacks. In fact, under such a parametrization, the number of model parameters increases amply as the $\rho_{i,t}$s are time-variant. This impacts on the computational procedures of the likelihood function.

5.2   Proposed Novelties

Taking into consideration the complications involved in the estimation and computational procedures in the above existing periodic INAR process, we propose at a first instance, a more straightforward high-ordered INAR process based on the simple binomial thinning procedure with constant coefficient and where the error or innovation term is the random component that accounts for the periodicity by allowing its predictor function to incorporate harmonic expression of the form

$$A\sin(2\pi\omega t) + B\cos(2\pi\omega t) \tag{5.2}$$

where $\omega > 0$. At the same time, the underlying error distribution can consider any discrete probability model with their zero-inflated associates to handle any form of over-dispersion in the data. Under these assumptions, it is expected that the proposed high-ordered INAR process is workable even though its conditional likelihood function may involve integrals. In fact, as Joe [2019] argued that with changed thinning mechanism such as the random or generalized binomial thinning, the computational performance of the conditional likelihood function is perturbed. Here, we aim at developing a high-ordered INAR model with simple constant binomial thinning under common innovation distributions with the harmonic structure that depend on the nature of the data. In the event the series is over-dispersed, the Poisson-Gamma mixtures or the marginal NB is considered since it proven lately in Chutoo et al. [2021] and also in Chapters 3 and 4 of this book, that for over-dispersed counts, for instance for the case of COVID-19 new infection cases for Mauritius, the NB or its zero-inflated version yield lower fitting criterion like the Akaike Information criterion. These models are also applicable if the series consist of huge number of zeros. In this book, in addition to the ZI version, we consider the hurdle extensions as well.

Note that, the Equation 5.2 is analogous to the definition of periodicity in the *ptest* package in R, with ω as the periodic constant. More on the extension of INAR (1) to INAR(p) processes under the binomial thinning procedure is already given at Section 2.2 of the Chapter 2.

5.3   Simulations

First, we consider an INAR (4) process with Poisson innovations as: $Y_t = \rho_1 * Y_{t-1} + \rho_2 * Y_{t-2} + \rho_3 * Y_{t-3} + \rho_4 * Y_{t-4} + R_t$, where the $R_t \sim \text{Poisson}(\lambda_t)$ with $\lambda_t = \exp(2\sin 2\pi\omega t + 3\cos 2\pi\omega t)$, with $\omega > 0$. Assume that $\rho_1 = 0.2, \rho_2 = 0.1, \rho_3 = 0.1, \rho_4 = 0.05$ and $\omega = 0.01, 1.25, 3.50$. Using these values, we simulate 100, 500 and 1000 observations. The graphs and properties of these time series are shown below:

1. $T = 100$. Using the *set.seed(1234)*, the data summary for $\omega = 0.01$ gives a minimum value of 0 and a maximum of 72 with the mean and variance at 15.7 and 502.6. The data also consists of around 30% of zeros. The *ptestg* gives a p-value of less than 0.0001 and hence confirms the harmonic or periodic nature of the generated time series. Now with $\omega = 1.25$ and *set.seed(1235)*, we obtain the minimum value of 1 and maximum of 51, with no zero observations and has mean at 14.4 and variance 151.4. The periodicity test gives significant p-values. For $\omega = 3.50$ and *set.seed(1236)*, the minimum is 1 and maximum at 18 with mean 6.76 and variance 16.49 and significant periodic test.

2. $T = 500$. We repeat the same experiments with same values of the thinning coefficients and ω and stored in *set.seed(2234)*, *set.seed(2235)* and *set.seed(2236)*.

In the above simulated data at $T = 500$, the mean and variance at $\omega = 0.01$ are (14.6 and 450.4), at $\omega = 1.25$, are (12.4 and 69.4) and at $\omega = 3.50$, are (6.72 and 17.8). The periodic tests are all significant. It is noticeable that for the different combination of ω, the data are over-dispersed, and as ω increases, the level of periodicity also increases. For small ω, the Fisher-Index of dispersion becomes more larger.

Since this book is focused on COVID-19 analysis in countries like Mauritius, where there was a preponderance of zero COVID-19 new infection and death cases in the initial stages, we consider the zero-inflated and hurdle-Poisson versions in the simulations as these can mimic the actual trend of the COVID-19 in countries like Mauritius to a large extent. An elaborate explanation on zero-inflated models is available in Chapter 3. The hurdle models which have been proposed by Mullahy [1986] are another class of family of models that are flexible to fit count data with zero-inflation and consists of two parts [Heilbron, 1994]. The first part is the Bernoulli for the zero counts and the second part is the zero-truncated for the positive counts [Mullahy, 1986, Min and Agresti, 2005, Rose et al., 2006, Zuur et al., 2009, Desjardins, 2013, 2016]. The hurdle model is usually expressed as:

$$P(R_t = r_t, \mu_t, v) = \begin{cases} \pi, & y_t = 0 \\ (1 - \pi), & \frac{P(R_t = r_t, \mu_t, v)}{1 - P(R_t = 0, \mu_t, v)}, \quad y_t > 0 \end{cases}$$

The PGF of the Hurdle-Poisson is given as $G_{R(t)}(s) = \pi + (1 - \pi)\left[\frac{\exp(\lambda_t(s-1))}{\exp(\lambda_t)-1}\right]$ and that of Hurdle-Negative Binomial will simply be written as

$$G_{R_t}(s) = \pi + (1 - \pi)\left[\left(\frac{\frac{1}{v}}{\mu_t + \frac{1}{v}}\right)^{\frac{1}{v}}\left[1 - \left(\frac{\frac{1}{v}}{\mu_t + \frac{1}{v}}\right)^{\frac{1}{v}}\right]^{-1}\left[\frac{(\mu_t + \frac{1}{v})^{\frac{1}{v}}}{(\mu_t + \frac{1}{v} - \mu_t s)^{\frac{1}{v}}} - 1\right]\right].$$

Using the earlier mentioned values for ρ and ω, an attempt was made to generate the periodic series under three different scenarios namely:

1. R_t is Poisson with mean λ_t

2. R_t is Zero-inflated Poisson (ZI-P) with mean λ_t and with the proportion of zeros $\pi = 0.4$. In this case, we use the *ZIM* package to obtain simulated zero-inflated Poisson data using the built in function *rzip*. For the ZI-P, we assume that the PGF of the ZI-P is given in Chapter 3.

3. R_t is Hurdle-Poisson with $\pi = 0.4$. Here we refer to the *iZID* package using the *sample.h* function.

In cases 2 and 3, η is thus computed as -0.4055. The simulation results based on 500 replications are tabulated in Tables 5.1-5.3. Below, the boxplot representing the simulated mean estimates under Poisson innovation is given: The boxplot with the simulated mean estimates under ZI-Poisson innovation is given in Figure 5.1. The boxplot with the simulated mean estimates under Hurdle-Poisson innovation is given in Figure 5.2. From the simulated results in Tables 5.1 to 5.3, it can be concluded that

Model	T	$\rho_1=0.2$	$\rho_2=0.1$	$\rho_3=0.1$	$\rho_4=0.05$	$\omega=0.01$	$\eta=-0.4055$
Poisson	100	0.215	0.101	0.111	0.048	0.014	−0.3956
		(0.016)	(0.015)	(0.018)	(0.009)	(0.019)	(0.005)
	500	0.204	0.095	0.128	0.053	0.013	−0.4025
		(0.015)	(0.013)	(0.014)	(0.008)	(0.016)	(0.003)
	1000	0.189	0.101	0.146	0.046	0.009	−0.4061
		(0.011)	(0.009)	(0.012)	(0.006)	(0.011)	(0.001)
ZI-P	100	0.208	0.099	0.137	0.053	0.011	−0.4056
		(0.009)	(0.006)	(0.012)	(0.009)	(0.013)	(0.009)
	500	0.215	0.145	0.122	0.05	0.010	−0.3949
		(0.005)	(0.004)	(0.008)	(0.005)	(0.008)	(0.005)
	1000	0.186	0.111	0.105	0.054	0.012	−0.4022
		(0.003)	(0.002)	(0.005)	(0.002)	(0.006)	(0.003)
HP	100	0.198	0.112	0.077	0.048	0.013	−0.410
		(0.005)	(0.006)	(0.010)	(0.008)	(0.009)	(0.006)
	500	0.220	0.149	0.105	0.055	0.011	−0.389
		(0.003)	(0.004)	(0.009)	(0.006)	(0.008)	(0.004)
	1000	0.218	0.116	0.142	0.053	0.014	−0.406
		(0.002)	(0.002)	(0.007)	(0.004)	(0.004)	(0.003)

TABLE 5.1
Simulated mean estimates for $\omega = 0.01$

Model	T	$\rho_1=0.2$	$\rho_2=0.1$	$\rho_3=0.1$	$\rho_4=0.05$	$\omega=1.25$	$\eta=-0.4055$
Poisson	100	0.199	0.126	0.112	0.044	1.205	−0.4052
		(0.021)	(0.016)	(0.019)	(0.025)	(0.011)	(0.006)
	500	0.214	0.117	0.106	0.051	1.196	−0.3998
		(0.018)	(0.013)	(0.013)	(0.018)	(0.009)	(0.004)
	1000	0.208	0.089	0.078	0.049	1.239	−0.4059
		(0.016)	(0.009)	(0.010)	(0.015)	(0.007)	(0.002)
ZI-P	100	0.238	0.069	0.096	0.043	1.244	−0.4023
		(0.019)	(0.013)	(0.012)	(0.014)	(0.019)	(0.009)
	500	0.201	0.112	0.099	0.058	1.231	−0.3965
		(0.012)	(0.006)	(0.010)	(0.006)	(0.018)	(0.005)
	1000	0.222	0.126	0.089	0.051	1.249	−0.4019
		(0.008)	(0.005)	(0.008)	(0.005)	(0.015)	(0.001)
HP	100	0.192	0.119	0.115	0.052	1.223	−0.4046
		(0.015)	(0.011)	(0.005)	(0.008)	(0.013)	(0.004)
	500	0.204	0.122	0.134	0.046	1.251	−0.4056
		(0.009)	(0.007)	(0.003)	(0.006)	(0.009)	(0.003)
	1000	0.239	0.102	0.144	0.045	1.196	−0.4002
		(0.008)	(0.004)	(0.002)	(0.005)	(0.008)	(0.001)

TABLE 5.2
Simulated mean estimates for $\omega = 1.25$

Model	T	$\rho_1=0.2$	$\rho_2=0.1$	$\rho_3=0.1$	$\rho_4=0.05$	$\omega=3.50$	$\eta=-0.4055$
Poisson	100	0.177	0.125	0.067	0.053	3.489	−0.410
		(0.015)	(0.020)	(0.026)	(0.012)	(0.019)	(0.021)
	500	0.191	0.108	0.089	0.046	3.523	−0.406
		(0.011)	(0.015)	(0.022)	(0.009)	(0.017)	(0.019)
	1000	0.224	0.114	0.123	0.049	3.489	−0.390
		(0.008)	(0.011)	(0.019)	(0.005)	(0.013)	(0.015)
ZI-P	100	0.207	0.142	0.107	0.050	3.556	−0.406
		(0.012)	(0.019)	(0.021)	(0.009)	(0.016)	(0.017)
	500	0.176	0.134	0.131	0.057	3.516	−0.405
		(0.009)	(0.015)	(0.016)	(0.007)	(0.011)	(0.013)
	1000	0.221	0.066	0.127	0.055	3.489	−0.403
		(0.007)	(0.008)	(0.014)	(0.006)	(0.009)	(0.011)
HP	100	0.179	0.112	0.133	0.049	3.467	−0.400
		(0.019)	(0.011)	(0.008)	(0.016)	(0.012)	(0.009)
	500	0.219	0.096	0.149	0.045	3.562	−0.400
		(0.012)	(0.007)	(0.006)	(0.009)	(0.009)	(0.006)
	1000	0.237	0.089	0.106	0.052	3.502	−0.403
		(0.008)	(0.005)	(0.004)	(0.006)	(0.005)	(0.004)

TABLE 5.3
Simulated mean estimates for $\omega = 3.50$

the simulated mean estimates of the model parameters on the 500 replications, in particular for the ρ and η are consistent with their corresponding population parameters. In fact, by comparing the biases, we note that the periodic INAR (4) with ZI-P innovation distribution model provides estimates with lower standard errors than the

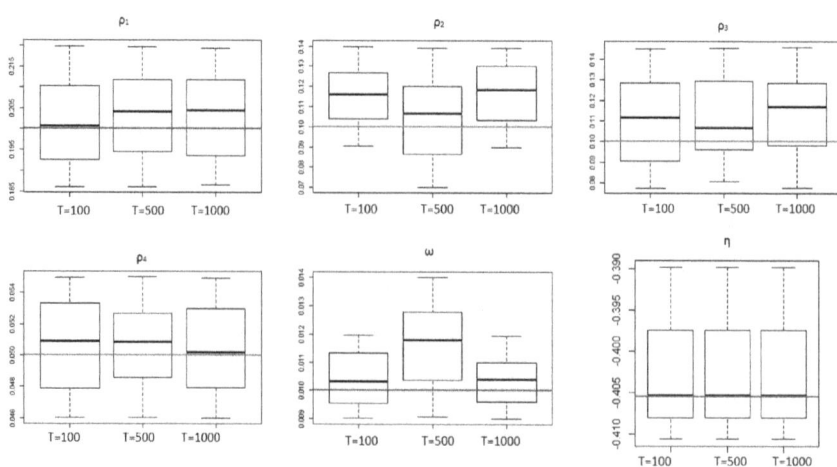

FIGURE 5.1
Boxplot for simulated Poisson innovation with $T = 100, \omega = 0.01$

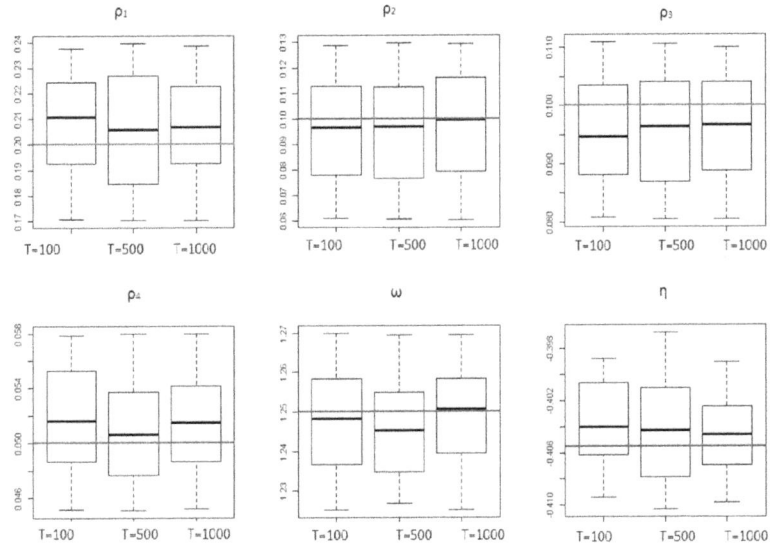

FIGURE 5.2
Boxplot for simulated ZI-P innovation with $T = 500, \omega = 1.25$

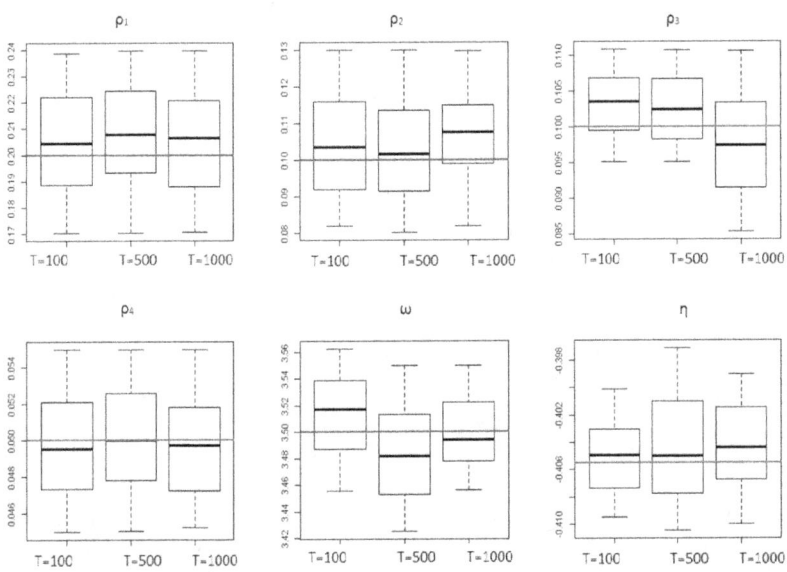

FIGURE 5.3
Boxplot for simulated Hurdle-Poisson innovation with $T = 1000, \omega = 3.50$

other competitive INAR (4) under stationary scenario. In fact, as the number of time points increases, the standard errors of the estimates decrease thus we can say that the consistency is unaffected by the ω. Also, computationally, simulations at T=500 and T=1000 are very time consuming. We also attempt higher order INAR with $p > 4$ and notice that the computational procedures become very slow and time consuming.

5.4   Data Analysis

5.4.1   The COVID-19 New Infection and Death Series in South Africa

Referring to Figure 1.3 in Chapter 1 and a significant periodicity test namely the *ptestg* test from the *ptest* package in R, it can be concluded that the COVID-19 new infection and death cases series of South Africa from 05 March 2020 to 22 November 2021, is highly autocorrelated, over-dispersed, high-ordered, non-stationary, of higher integers and most importantly, has periodicity. Therefore, we start by fitting 5 models based on order=4 due to the complexities involved in running the codes and also to adhere to the rule of parsimony, which are, the INAR (4) with Poisson innovations (Model 1), the periodic INAR (4) with Poisson innovations (Model 2), the periodic INAR (4) with Negative Binomial innovations (Model 3), the periodic INAR (4) with Poisson Weighted-Exponential innovation (Model 4) and the periodic INAR (4) with Poisson Generalized-Lindley distribution innovation (Model 5), to the COVID-19 new infection series of South Africa. Note that the purpose of fitting using Model 1 is to reflect on the biasedness that may occur when we ignore the periodicity in the data analysis. The estimates of the parameters under these models and their standard errors and AIC are tabulated in Table 5.4. Based on the results of Table 5.4, we observed that all parameters estimates are significant which means that periodic high-ordered INAR processes suit this series well. However, based on the AIC criterion, it is observed that periodic INAR (4) with NB (Model 3) yield the lowest AIC, followed by periodic INAR (4) PGLD (Model 5). It is noteworthy that Model 1, INAR (4) with Poisson innovations, which did not account for periodicity, yield comparatively less reliable fits. Thus, it is important to cater for harmonic or periodic functions when working with COVID-19 new infection series for South Africa. Next, a similar exercise is conducted on the COVID-19 new death series; however, here we also assumed an order, p, of 4. As shown in Table 5.5, the parameter estimates under all models for the COVID-19 new death series are all significant with the p-values less than 0.05. Based on the AIC results above, Model 3 which is periodic INAR (4) with Negative Binomial innovation followed this time by periodic INAR (4) PWE innovation, again yield lower AIC. Here, it can be deduced that the harmonic specification in the definition of the link predictor function λ_t plays an important role as Model 2 to Model 5 had better fits than Model 1 (Recall: Model 1 did not consider any periodic feature).

Models	Parameters	ρ_1	ρ_2	ρ_3	ρ_4	λ	ω	AIC
Model 1	Estimates	0.389	0.196	0.268	0.053	0.496		17449
	StdError	(0.001)	(0.000)	(0.000)	(0.000)	(0.001)		
Model 2	Estimates	0.259	0.241	0.196	0.092		0.021	16586
	StdError	(0.000)	(0.000)	(0.000)	(0.000)		(0.000)	
Model 3	Estimates	0.502	0.203	0.217	0.076		0.015	15066
	StdError	(0.000)	(0.000)	(0.000)	(0.000)		(0.000)	
Model 4	Estimates	0.434	0.198	0.223	0.089		0.012	16024
	StdError	(0.000)	(0.000)	(0.000)	(0.000)		(0.000)	
Model 5	Estimates	0.289	0.216	0.227	0.082		0.018	15343
	StdError	(0.000)	(0.000)	(0.000)	(0.000)		(0.000)	

TABLE 5.4
Estimates of the model parameters under different INAR (4) processes: COVID-19
new infection series of South Africa

Models	Parameters	ρ_1	ρ_2	ρ_3	ρ_4	λ	ω	AIC
Model 1	Estimates	0.426	0.170	0.194	0.030	0.439		14523
	StdError	(0.000)	(0.000)	(0.000)	(0.000)	(0.000)		
Model 2	Estimates	0.263	0.227	0.214	0.084		0.016	14503
	StdError	(0.000)	(0.000)	(0.000)	(0.000)		(0.000)	
Model 3	Estimates	0.455	0.211	0.224	0.082		0.015	14489
	StdError	(0.000)	(0.000)	(0.000)	(0.000)		(0.000)	
Model 4	Estimates	0.410	0.216	0.205	0.091		0.012	14497
	StdError	(0.000)	(0.000)	(0.000)	(0.000)		(0.000)	
Model 5	Estimates	0.354	0.198	0.232	0.084		0.018	14501
	StdError	(0.000)	(0.000)	(0.000)	(0.000)		(0.000)	

TABLE 5.5
Estimates of the model parameters under different INAR (4) processes: COVID-19
new death series of South Africa

As an addition, for the COVID-19 new death series which had least been studied, we consider the effects of some influential time-varying factors such as the
COVID-19 Stringency Index, and the number of new COVID-19 infection cases.
It is important to study the death series given that it gives an indication of the severity of the disease as well as the effectiveness of the remedial actions and sanitary
related policies.

We start by re-formulating the λ_t as follows:

$$\lambda_t = \exp(\beta_0 \sin 2\pi\omega t + \beta_1 \cos 2\pi\omega t) \times \text{infected}^{\beta_2} \times \text{stringency}^{\beta_3} \quad (5.3)$$

Models	Parameters	ρ_1	ρ_2	ρ_3	ρ_4	β_0	β_1	β_2	β_3	ω	AIC
Model 1	Estimates	0.459	0.170	0.179	0.020			0.009	0.008		14352
	StdError	(0.000)	(0.000)	(0.000)	(0.002)			(0.000)	(0.000)		
Model 2	Estimates	0.453	0.156	0.184	0.030	0.008	0.006	0.015	0.006	0.011	14246
	StdError	(0.000)	(0.000)	(0.000)	(0.000)	(0.003)	(0.000)	(0.000)	(0.000)	(0.000)	
Model 3	Estimates	0.443	0.160	0.186	0.025	0.008	0.006	0.015	0.005	0.011	14219
	StdError	(0.000)	(0.000)	(0.000)	(0.000)	(0.000)	(0.000)	(0.000)	(0.000)	(0.000)	
Model 4	Estimates	0.402	0.150	0.195	0.024	0.007	0.005	0.016	0.004	0.012	14235
	StdError	(0.000)	(0.000)	(0.000)	(0.000)	(0.000)	(0.000)	(0.000)	(0.000)	(0.000)	
Model 5	Estimates	0.417	0.163	0.179	0.019	0.008	0.009	0.015	0.006	0.013	14229
	StdError	(0.000)	(0.000)	(0.000)	(0.000)	(0.000)	(0.000)	(0.000)	(0.000)	(0.000)	

TABLE 5.6
Estimates of the model parameters under different INAR (4) with covariates: South Africa

As mentioned in Chapter 3, recall that the COVID-19 Stringency Index gives an indication of the strictness of the sanitary restrictions imposed in relation to the COVID-19 pandemic and the higher the value of the index, the stricter are the sanitary measures. In Table 5.6, all Models 1 to 5 yield reliable estimates of the different parameters at very low standard errors and hence with significant p-values. However, Model 3 still gives comparatively better AIC that is periodicity was catered better by the periodic INAR (4) with Negative Binomial model, followed by the INAR (4) with PGLD and PWE. Based on the highly significant estimates of the COVID-19 stringency index, it can be deduced that indeed the implementation of sanitary measures in South Africa like stringent lockdown, limited mobility and gatherings and legislation such as the Disaster Management Act as part of its preparedness plan, allowed South African authorities to timely detect variants and come up with proper remedial actions to strengthen the health system. Consequently, a higher COVID-19 Stringency index implies stricter sanitary restrictions thus better containing the propagation of the SARS-CoV-2 and its variants. As for the COVID-19 new infection cases variable which is highly significant, it can be deduced that when the number of COVID-19 related contamination cases is higher, the number of associated death cases will rise as well.

We now shift to another SIDS which had gain much visibility given that it was amongst the first country to have curbed the COVID-19 pandemic in 2020 via its strict sanitary restrictions—Mauritius [Mamode Khan et al., 2020a].

5.4.2 COVID-19 New Infection and Deaths in Mauritius

After a thorough analysis of the COVID-19 new infection series in Chapter 3, we again attempt to assess this series and the related new COVID-19 death cases of

Models	Parameters	ρ_1	ρ_2	ρ_3	ρ_4	β_0	β_1	β_2	β_3	ω	AIC
Model 1	Estimates	0.189	0.098	0.126	0.049			0.019	0.009		31245
	StdError	(0.002)	(0.003)	(0.007)	(0.004)			(0.004)	(0.002)		
Model 2	Estimates	0.102	0.102	0.099	0.055	0.023	0.022	0.011	0.014	0.018	28965
	StdError	(0.005)	(0.009)	(0.011)	(0.011)	(0.003)	(0.000)	(0.004)	(0.003)	(0.001)	
Model 3	Estimates	0.243	0.089	0.125	0.045	0.012	0.016	0.010	0.029	0.013	25106
	StdError	(0.008)	(0.004)	(0.006)	(0.007)	(0.001)	(0.004)	(0.003)	(0.003)	(0.005)	
Model 4	Estimates	0.256	0.079	0.136	0.051	0.018	0.028	0.009	0.013	0.024	26052
	StdError	(0.010)	(0.004)	(0.002)	(0.004)	(0.001)	(0.004)	(0.001)	(0.003)	(0.003)	
Model 5	Estimates	0.205	0.112	0.133	0.054	0.012	0.021	0.017	0.009	0.031	25324
	StdError	(0.009)	(0.005)	(0.001)	(0.004)	(0.006)	(0.001)	(0.001)	(0.003)	(0.002)	
Model 6	Estimates	0.195	0.106	0.142	0.063	0.019	0.034	0.024	0.010	0.039	28302
	StdError	(0.001)	(0.003)	(0.001)	(0.002)	(0.005)	(0.001)	(0.001)	(0.000)	(0.002)	
Model 7	Estimates	0.242	0.119	0.147	0.059	0.011	0.019	0.021	0.009	0.042	27865
	StdError	(0.003)	(0.005)	(0.001)	(0.001)	(0.003)	(0.001)	(0.001)	(0.003)	(0.001)	

TABLE 5.7

Estimates of the model parameters under different INAR (4) processes: COVID-19 new infection series in Mauritius

Mauritius, but this time, by catering for the periodic feature under high ordered INAR processes with some Poisson-mixtures innovation distribution and their ZI versions. Here, the Mauritian' COVID-19 new infection and death series from 18 March 2020 to 25 April 2021 was used for this analysis and we test the INAR model with Poisson model (Model 1: No periodic), the periodic INAR (4) model with zero-inflated Poisson (ZI-P) (Model 2 : ZI-P), the periodic INAR (4) model with zero-inflated Negative-Binomial innovations (Model 3: ZI-NB), the periodic INAR (4) model with zero-inflated PWE innovations (Model 4: ZI-PWE), the periodic INAR (4) model with zero-inflated PGLD innovations (Model 5: ZI-PGLD) and even introduce the periodic INAR (4) model with Hurdle Poisson innovation (Model 6: H-P) and the periodic INAR (4) model with Hurdle Negative-Binomial (Model 7:H-NB). In fact, both COVID-19 series for Mauritius contain an excess of zeros, that is around 97.3% of zero occurrences, as compared to the South African COVID-19 data thus accounting for hurdle models here, because of its flexibility to account for excess zeros, will give another dimension to this analysis. The results are shown in Tables 5.7-5.8.

From the results of Tables 5.7 and 5.8, all models under both series, yield significant estimates with low *p*-values. Also, for both the COVID-19 new infection cases and for the COVID-19 new death series for Mauritius, the periodic INAR (4) model with zero-inflated Negative-Binomial innovations (Model 3: ZI-NB), outperformed the other ZI and Hurdle models, that is, had the lowest AIC. In both scenarios, the INAR processes wiht ZI-PGLD and ZI-PWE innovations provided satisfactorily

Models	Parameters	ρ_1	ρ_2	ρ_3	ρ_4	β_0	β_1	β_2	β_3	ω	AIC
Model 1	Estimates	0.213	0.107	0.107	0.054			0.011	0.008		25030
	StdError	(0.004)	(0.006)	(0.009)	(0.005)			(0.004)	(0.007)		
Model 2	Estimates	0.142	0.093	0.107	0.057	0.015	0.012	0.013	0.015	0.004	25024
	StdError	(0.011)	(0.019)	(0.010)	(0.013)	(0.003)	(0.006)	(0.005)	(0.003)	(0.013)	
Model 3	Estimates	0.165	0.086	0.103	0.050	0.009	0.011	0.010	0.011	0.012	20080
	StdError	(0.010)	(0.007)	(0.006)	(0.007)	(0.006)	(0.004)	(0.003)	(0.003)	(0.009)	
Model 4	Estimates	0.169	0.086	0.107	0.053	0.009	0.018	0.011	0.012	0.019	22120
	StdError	(0.012)	(0.008)	(0.003)	(0.007)	(0.006)	(0.004)	(0.003)	(0.003)	(0.006)	
Model 5	Estimates	0.149	0.089	0.116	0.048	0.009	0.015	0.014	0.012	0.016	23412
	StdError	(0.010)	(0.007)	(0.002)	(0.004)	(0.006)	(0.004)	(0.003)	(0.003)	(0.008)	
Model 6	Estimates	0.153	0.092	0.124	0.053	0.012	0.016	0.013	0.014	0.023	25102
	StdError	(0.009)	(0.006)	(0.002)	(0.003)	(0.005)	(0.003)	(0.001)	(0.003)	(0.005)	
Model 7	Estimates	0.155	0.094	0.119	0.049	0.009	0.014	0.009	0.012	0.019	24029
	StdError	(0.012)	(0.005)	(0.005)	(0.006)	(0.005)	(0.005)	(0.003)	(0.005)	(0.009)	

TABLE 5.8
Estimates of the model parameters under different INAR (4) processes: COVID-19 new death series in Mauritius

results thus required further exploration, possibly under other research areas. On the other hand, ignoring the computational failures of the Hurdle models, we can observe that INAR (4) with Hurdle-NB innovations also show satisfactory AIC.

We again worked with the COVID-19 new death series for Mauritius. All the above models have been explored while catering for influential factors, that is COVID-19 Stringency Index and the daily count of COVID-19 new infection cases. Interesting results are shown in Table 5.9: The Model 3, the periodic INAR (4) with ZI-Negative Binomial innovations, still reaps the lowest AIC in comparison to the other ZI and hurdle models with an over-dispersion parameter of 0.428 and standard errors of 0.002. However, in this exercise, the computational complexities involved were huge as the operation was very time consuming. Note that the COVID-19 Stringency Index was significant, implying that the timely imposition of new immediate sanitary measures during the peak COVID-19 phases, like sanitary curfew/lockdown, sanitization and sensitization campaigns, and safe shopping guidelines, played a vital role in reducing the number of deaths related to COVID-19. Even the new cases of COVID-19 infection are reported as significant, mainly due to the close causal relationship, as can be seen visually in Figure 1.6 in Chapter 1. This aspect however necessitate more exploration possibly via a bivariate model.

Models	Parameters	ρ_1	ρ_2	ρ_3	ρ_4	β_0	β_1	β_2	β_3	ω	η	AIC
Model 1	Estimates	0.144	0.112	0.109	0.052			0.011	0.012		0.921	25025
	StdError	(0.005)	(0.010)	(0.010)	(0.007)			(0.003)	(0.002)		(0.004)	
Model 2	Estimates	0.141	0.111	0.106	0.051	0.011	0.011	0.011	0.012	0.011	0.931	25019
	StdError	(0.009)	(0.006)	(0.004)	(0.006)	(0.003)	(0.003)	(0.002)	(0.004)	(0.002)	(0.003)	
Model 3	Estimates	0.175	0.098	0.091	0.049	0.011	0.011	0.008	0.009	0.011	0.911	20078
	StdError	(0.007)	(0.005)	(0.003)	(0.004)	(0.002)	(0.003)	(0.002)	(0.003)	(0.002)	(0.002)	
Model 4	Estimates	0.156	0.122	0.103	0.048	0.014	0.012	0.008	0.009	0.009	0.896	20942
	StdError	(0.006)	(0.003)	(0.002)	(0.004)	(0.003)	(0.002)	(0.001)	(0.003)	(0.001)	(0.000)	
Model 5	Estimates	0.145	0.115	0.098	0.054	0.010	0.013	0.009	0.011	0.012	0.913	20515
	StdError	(0.005)	(0.003)	(0.001)	(0.003)	(0.002)	(0.001)	(0.000)	(0.002)	(0.002)	(0.003)	
Model 6	Estimates	0.146	0.112	0.105	0.052	0.011	0.011	0.011	0.012	0.010	0.012	25020
	StdError	(0.010)	(0.007)	(0.004)	(0.005)	(0.004)	(0.003)	(0.002)	(0.003)	(0.002)	(0.003)	
Model 7	Estimates	0.149	0.109	0.123	0.052	0.011	0.009	0.010	0.012	0.010	0.013	21010
	StdError	(0.012)	(0.005)	(0.003)	(0.005)	(0.004)	(0.003)	(0.001)	(0.003)	(0.001)	(0.004)	

TABLE 5.9

Estimates of Parameters of INAR (4) with different ZI and Hurdle innovations with covariates: Mauritius

5.4.3 Forecasts

An outsample forecast for the COVID-19 new infection cases was conducted for South Africa covering the period from 23 November 2021 to 02 December 2021 using the Model 3 (periodic INAR (4) with NB innovations), while for Mauritius, the forecasting period for the COVID-19 new infection cases starts from 26 April 2021 to 05 May 2021 using the Model 3 (periodic INAR (4) with ZI-NB innovations), as shown in Figure 5.4: The forecasts reaped satisfactory RMSEs notably for South Africa which was around 5.29 whilst for Mauritius, around 1.73. The confidence interval is shown in Figure 5.5.

5.4.4 Proposed Extension: A Novel Bivariate INAR (1) with Harmonic Functions Model

As depicted in Figure 1.6 under Chapter 1, a strong inter-relationship between the COVID-19 new infection and death series, was observed. Thus, we move forward in this work and propose a periodic bivariate INAR (1) model under the binomial thinning operator mainly because, its execution is more smooth compared to the other GB and NB thinning, as reported earlier in Chapters 3 and 4. We still focus

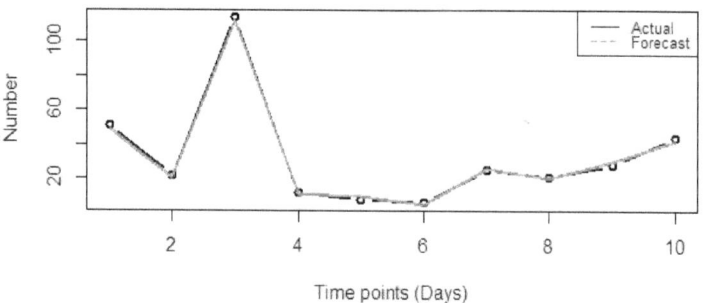

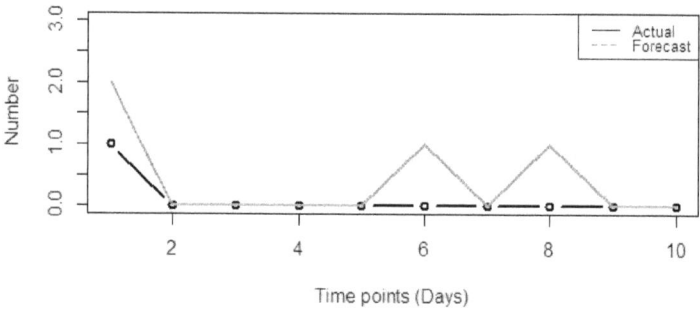

FIGURE 5.4

Forecasts for next 10 days

on the COVID-19 new infection and new death cases of Mauritius. Below are the formulations:

$$Y_{t,1} = \rho_{11} * Y_{t-1,1} + R_{t,1}$$
$$Y_{t,2} = \rho_{21} * Y_{t-1,2} + \rho_{22} * Y_{t-1,1} + R_{t,2},$$

where $Y_{t,1}$ measures the number of COVID-19 new infection cases at the tth time point and $Y_{t,2}$ is the corresponding number of COVID-19 new death cases. The covariance between $Y_{t,1}$ and $Y_{t,2}$ is given by: $Cov(Y_{t,1}, Y_{t,2}) = Cov(\rho_{11} * Y_{t-1,1}, \rho_{21} * Y_{t-1,2}) + Cov(\rho_{11} * Y_{t-1,1}, \rho_{22} * Y_{t-1,1})$, where $Cov(R_{t,1}, R_{t,2}) = 0$. The $*$ operator is the usual binomial thinning operator. The assumption on the previous lagged term $Y_{t-j,k}$ and $R_{t,k}$ holds as in the classical INAR time series model. The Bernoulli sequences in the paired terms in $Y_{t,2}$ that is $(\rho_{21} * Y_{t-1,2}, \rho_{22} * Y_{t-1,1})$ is treated independent.

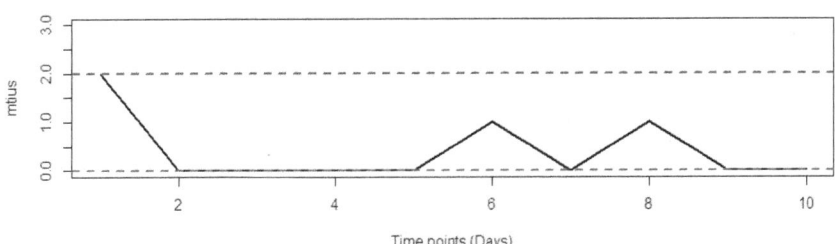

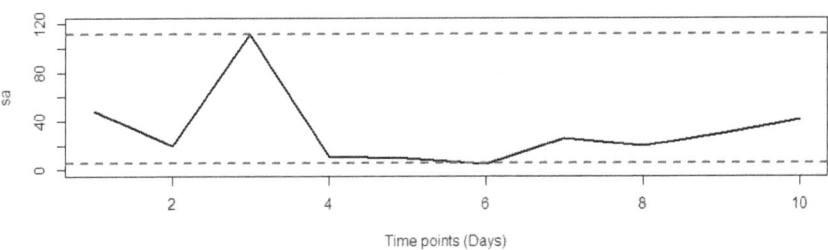

FIGURE 5.5
The forecasts within 95% CI

We further assume that R_{t-1} and $R_{t,2}$ follow the ZI-NB, ZI-PWE and ZI-PGLD mainly due to the excess of zeros and over-dispersion features observed in both series. Note, given the ZI models, we assume the probability of zeros denoted by π_j, where $j = 1, 2$. The choice of the ZI models for both $R_{t,j}$ is because the series $Y_{t,1}$ and $Y_{t,2}$ are hugely over-dispersed, and are characterized by a large number of zeros. The marginal means of $R_{t,1}$ and $R_{t,2}$ are thus given as:

$$\lambda_{t,1} = \exp(\beta_{10} \sin 2\pi\omega_1 t + \beta_{11} \cos 2\pi\omega_1 t) \times \text{stringency}^{\beta_{12}}$$

$$\lambda_{t,2} = \exp(\beta_{20} \sin 2\pi\omega_2 t + \beta_{21} \cos 2\pi\omega_2 t) \times \text{stringency}^{\beta_{22}}$$

where 'stringency' refers to the covariate—'COVID-19 Stringency Index'. As already mentioned in Chapters 3 and 4, the Fisher Dispersion Index (FI) for these models have been found to be well suited to model over-dispersed series. In the above equation, we take the COVID-19 Stringency Index as a time-varying covariate to induce the non-stationarity in both series.

5.4.5 Application to COVID-19 New Infection and Death Cases in Mauritius

Still focusing on Mauritius, since its COVID-19 new infection and death cases did not consist of huge values, we start by applying the proposed novel periodic BINAR

Model	Parameters	$\beta_1 0$	$\beta_1 1$	$\beta_1 2$	$\beta_2 0$	$\beta_2 1$	$\beta_2 2$	π_1	ω	AIC	Log-likelihood
ZI-NB	Estimates	0.329	0.124	0.289	0.395	0.175	0.241	0.896	0.126	1806.5	1792.5
	Std Errors	(0.000)	(0.002)	(0.000)	(0.001)	(0.001)	(0.000)	(0.001)	(0.000)		
ZI-PWE	Estimates	0.308	0.098	0.349	0.426	0.223	0.289	0.915	0.116	1912.3	1898.3
	Std Errors	(0.001)	(0.000)	(0.002)	(0.000)	(0.000)	(0.002)	(0.000)	(0.000)		
ZI-PGLD	Estimates	0.276	0.115	0.267	0.389	0.196	0.315	0.934	0.103	1897.6	1883.6
	Std Errors	(0.002)	(0.001)	(0.001)	(0.000)	(0.002)	(0.000)	(0.000)	(0.000)		

TABLE 5.10

Estimates of the BINAR (1) process with different ZI innovations : COVID-19 new infection series

Model	Parameters	$\beta_1 0$	$\beta_1 1$	$\beta_1 2$	$\beta_2 0$	$\beta_2 1$	$\beta_2 2$	π_1	ω	AIC	Log-likelihood
ZI-NB	Estimates	0.296	0.121	0.326	0.405	0.213	0.232	0.923	0.096	3436.7	3422.7
	Std Errors	(0.002)	(0.001)	(0.000)	(0.001)	(0.000)	(0.000)	(0.000)	(0.000)		
ZI-PWE	Estimates	0.365	0.093	0.339	0.436	0.196	0.196	0.869	0.136	5965.3	5951.3
	Std Errors	(0.003)	(0.000)	(0.000)	(0.003)	(0.001)	(0.001)	(0.002)	(0.001)		
ZI-PGLD	Estimates	0.289	0.103	0.276	0.396	0.206	0.218	0.931	0.112	6126.9	6112.9
	Std Errors	(0.002)	(0.001)	(0.001)	(0.002)	(0.000)	(0.002)	(0.001)	(0.000)		

TABLE 5.11

Estimates of the BINAR (1) process with different ZI innovations: COVID-19 new death series

(1) to the COVID-19 new infection and death series under the ZI-NB, ZI-PWE, and ZI-PGLD models. The results are summarized in Tables 5.10 and 5.11.

From Tables 5.10 and 5.11, it can be obtained that under the COVID-19 new infection and death series, the periodic BINAR (1) with ZI-NB innovation, with comparatively lower AIC, performed better. Here, it is important again to highlight the computational complexities in running the codes. However, conclusively, the results are reliable with significant COVID-19 Stringency index.

5.5　Concluding Remarks

This particular work brings an important finding in the class of integer-valued autoregressive models. For series with harmonic or periodic structure, it is important to consider the periodic feature with some harmonic functions in the definition of the innovation predictor function whilst, the choice of the error term is determined by the level of over-dispersion in the data and the presence of zeros. This subsequently impacts on the significance of the explanatory variables and obviously the fitting criteria of the time series models. In the presence of over-dispersion and excess zeros, we considered different Poisson-mixtures models in the error term specification, depending on the data application. For instance, for the COVID-19 series for South Africa, the INAR (4) conventional Poisson-mixtures were used under the binomial thinning, whilst for Mauritius, since its COVID-19 new infection and death series consisted of large number of zero cases, we even added the ZI and Hurdle versions of the Poisson-mixtures model. Conclusively, for South Africa, the periodic INAR (4) with Negative Binomial innovation was the best suited models whilst for Mauritius, the periodic INAR (4) with ZI-NB innovation models outperformed the other models. Note here, we wish to again highlight the computational complexities involved when working with the periodic and high-ordered INAR algorithm especially with novel models like PWE and PGLD and their ZI-versions.

Furthermore, given the close relationship between the COVID-19 new infection and death series, a preliminary work on extending the periodic high ordered INAR models to a bivariate INAR model has also been done in this Chapter 5. The results for both series are again in favor of BINAR (1) with ZI-NB innovations. It would now be advantageous to further explore the PWE and PGLD, considering their satisfactory results.

5.6 Important Point to Highlight

It is important to highlight that in Chapter 3, when we were dealing with the COVID-19 new infection cases of Mauritius, the ZI-NB outperformed all models and in Chapter 4, when dealing with the COVID-19 new death series for South Africa, the periodic INAR (4) with NB innovations followed by PWE and PGLD are amongst the best models. For Mauritius, on the other hand, the periodic INAR (4) with ZI-NB innovations followed by PGLD and PWE are amongst the best models. These findings were impressive since the Negative Binomial or its zero-inflated associates seem to work well in any sort of data, be it on lower integers, over-dispersed, non-stationary, periodic, oscillated series, or on higher integers with mentioned additional features. It will now be more interesting to explore the other two best performing models that is the PWE and PGLD models and their ZI versions which are recently emerged models on possibly different data applications. In fact, in the subsequent Chapters, based on the CCF plots at Figure 1.6 and 1.7, and to understand the strong inter-relation between COVID-19 new infection cases and deaths cases, we shall move towards working on bivariate processes under the binomial thinning procedure. The latter is still preferred because, as observed during Chapters 3, 4, and 5, the binomial thinning is more executable. Also, when we moved to bivariate processes, the computational algorithm becomes more complex and time consuming. Thus, we start by extending the bivariate of order 1 process with PWE and PGLD first. Finally, it is known that the close relationship between the two series gives an indication of the severity of the pandemic thus allowing better policy decision especially in terms of revamping the hospital infrastructure (bed capacity), possible vaccination requirement and so on.

6

Exploring the Bivariate Processes—The Bivariate INAR (1) Model with Paired Poisson—Weighted Exponential Distributions

6.1 Introduction

The growing number of applications that involve time series of counts has urged for more suitable integer-valued time series models that can handle the commonest phenomenon of over-dispersion, while taking into consideration the instances where two related series are collected Pedeli and Karlis [2011], Ristic et al. [2012], Pedeli and Karlis [2013aa], Mamode Khan et al. [2016], Nastic et al. [2016], Sunechar et al. [2017], Jowaheer et al. [2017], Sunechar et al. [2018a], Mamode Khan et al. [2019]. Admittedly, in the literature, since the pioneered work of Steutel and Van Harn [1979], McKenzie [1986], Al Osh and Alzaid [1987], McKenzie [1988], Alzaid and Al Osh [1990], Al Osh and Aly [1992], there has been significant progress in the development of simple integer-valued auto-regressive time series models of order 1 (INAR (1)) with different innovation distributions Ristic et al. [2009], Bakouch and Ristic [2010], Schweer and Weiß [2014], Bourguignon and Vasconcellos [2015], Livio et al. [2018], Mohammadpour et al. [2018], Bakouch et al. [2021], Sharafi et al. [2021] and the references therein. These various INAR (1)s compete to provide the most superior fitting criteria. Until recently, Altun [2019] proposed an alternative INAR (1) with PWE, that is deemed suitable to model over-dispersed series while providing lower Akaike Information Criterion. In his work, Altun [2019] applied the developed INAR (1) with PWE innovations to the Shots and Drugs data from the Police Car beat in Pittsburgh, and noted that the INAR (1) PWE yielded better AICs than the INAR (1) with Poisson Lindley and Generalized Poisson distributions and this conclusion is rather motivating to further explore the PWE in other important settings such as in the BINAR case, and thereon compare with other bivariate INAR (1) processes as discussed in Pedeli and Karlis [2011] and assess its performance versus the other popular BINAR (1) models. In this process, we first review the properties of the PWE and its adaption to the INAR (1) process. From thereon to mount the bivariate INAR (1) process with PWE innovations, we propose to induce a cross correlation between the counting series by assuming the paired PWE innovations are jointly distributed. Up to this extent, there is yet no bivariate PWE distribution, and in this book, we therefore propose to construct flexible bivariate PWE model using

DOI: 10.1201/9781003677451-6

the approaches by Gomez Deniz et al. [2012] and Bermudez and Karlis [2021] that can account for both positive and negative cross correlation. The properties of the bivariate PWE models will be derived, followed by the BINAR (1) process where the paired innovations are assumed to follow the proposed bivariate PWE distributions. Since the bivariate PWE model is novel, we shall also present some Monte Carlo simulation results to assess the consistency of the bivariate PWE parameters. Estimation of the parameters will be done using the CML and CLS approaches and finally the application will be on real-life intra-day transaction series of some popular stocks and after comparing the new BINAR (1)s with PWE innovations with other popular BINAR (1) models, proper conclusion will be drawn.

6.1.1 Part A: The Bivariate Poisson-Weighted Exponential Distribution

As an extension of the work by Altun [2019], we consider to construct the bivariate PWE model using the approaches by Gomez Deniz et al. [2012]. To start with, we re-define the PWE distribution: the random variable Y has the Poisson Weighted Exponential (PWE) distribution if it satisfies the following stochastic representation,

$$Y \mid \mu_t \sim P(\mu_t),$$
$$\mu_t \mid \nu, \lambda_t \sim WE(\nu, \lambda_t),$$

where $\mu_t > 0$, $\lambda_t > 0$ and $\nu > 0$. The unconditional distribution of Y is called as a PWE distribution and the probability mass function (pmf) of it is given by

$$P(Y = y) = \nu(1 + \lambda_t)(1 + \nu + \nu\lambda_t)^{-y-1} \qquad y = 0, 1, 2, ... \qquad (6.1)$$

The random variable Y with density (6.1) is denoted by $Y \sim PWE(\nu, \lambda_t)$ and the probability generating function (pgf) and moment generating function (mgf) of it are given by

$$G(s) = \frac{\nu(\lambda_t + 1)}{1 - s + \nu(\lambda_t + 1)},$$

$$M(t) = \frac{\nu(\lambda_t + 1)}{1 - e^t + \nu(\lambda_t + 1)},$$

respectively. In the rest of this Chapter, we introduce two bivariate random variables based on PWE distribution and obtain their properties. A bivariate random variable $\mathbf{Y} = (Y_1, Y_2)^T$ has basic bivariate PWE (BPWE) distribution, if it satisfies the following stochastic representation

$$Y_i \mid \Lambda = \mu_t \sim P(\mu_t \phi_i),$$
$$\Lambda \sim NWE(\nu, \lambda_t),$$

where $\phi_i > 0$, $i = 1, 2$ and $Y_1 \mid \Lambda$ and $Y_2 \mid \Lambda$ are independent.

Proposition 1. *The pmf of the basic BPWE distribution is given by*

$$P(Y_1 = y_1, Y_2 = y_2) = v(1 + \lambda_t) \frac{(y_1 + y_2)!}{y_1! y_2!} \frac{\phi_1^{y_1} \phi_2^{y_2}}{\left(\phi_1 + \phi_2 + v(1 + \lambda_t)\right)^{y_1 + y_2 + 1}}, \quad (6.2)$$

where $y_1, y_2 = 0, 1, 2, \ldots$ and $\phi_1, \phi_2, v, \lambda_t > 0$.

Proof. According to Gomez Deniz et al. [2012] and by the Definition 2.1, the pmf of the basic BPWE distribution is

$$P(Y_1 = y_1, Y_2 = y_2) = \int_0^\infty P(Y_1 = y_1, Y_2 = y_2 \mid \Lambda = \mu_t) dF_{\mu_t}(\mu_t)$$

$$= \int_0^\infty P(Y_1 = y_1 \mid \Lambda = \mu_t) P(Y_2 = y_2 \mid \Lambda = \mu_t) F_\Lambda(\mu_t) d\mu_t$$

$$= \int_0^\infty \frac{(\mu_t \phi_1)^{y_1} e^{-\mu_t \phi_1}}{y_1} \frac{(\mu_t \phi_2)^{y_2} e^{-\mu_t \phi_2}}{y_2} v(1 + \lambda_t) e^{-\mu_t v(1 + \lambda_t)} d\mu_t$$

$$= v(1 + \lambda_t) \frac{\phi_1^{y_1} \phi_2^{y_2}}{y_1 y_2} \int_0^\infty \mu_t^{y_1 + y_2} e^{-(\phi_1 + \phi_2 + v(1 + \lambda_t))\mu_t} d\mu_t$$

$$= v(1 + \lambda_t) \frac{\phi_1^{y_1} \phi_2^{y_2}}{y_1 y_2} \frac{(y_1 + y_2)!}{(\phi_1 + \phi_2 + v(1 + \lambda_t))^{y_1 + y_2 + 1}}.$$

In the last equality, we used the density function of the Gamma distribution with parameters $(y_1 + y_2 + 1)$ and $(\phi_1 + \phi_2 + v(1 + \lambda_t))$. $\qquad \square$

The random vector $\mathbf{Y} = (Y_1, Y_2)^T$ with pmf (6.2) is denoted by $\mathbf{Y} \sim BPWE(v, \lambda_t, \phi_1, \phi_2)$.

Proposition 2. *Let $\mathbf{Y} \sim BPWE(v, \lambda_t, \phi_1, \phi_2)$, then*

a) *the marginal pmf of Y_i is* $P(Y_i = y_i) = \dfrac{v(1 + \lambda_t)}{\phi_i + v(1 + \lambda_t)} \left(\dfrac{\phi_i}{\phi_i + v(1 + \lambda_t)}\right)^{y_i}$, $y_i = 0, 1, 2, \ldots$ *i.e. for $i = 1, 2$, Y_i has the geometric distribution with parameter $\dfrac{v(1 + \lambda_t)}{\phi_i + v(1 + \lambda_t)}$.*

In special case, the $\phi_i = 1$ when $Y_i \sim PWE(v, \lambda_t)$ for $i = 1, 2$.

b) *The joint pgf of* $\mathbf{Y}$ *is given by* $G_Y(s_1,s_2) = \dfrac{v(1+\lambda_t)}{\phi_1(1-s_1)+\phi_2(1-s_2)+v(1+\lambda_t)}$,

where

$$0 < s_2 < \frac{\phi_1+\phi_2+v(1+\lambda_t)}{\phi_2} \text{ and } 0 < s_1 < \frac{\phi_1+\phi_2(1-s_2)+v(1+\lambda_t)}{\phi_1}.$$

c) *The conditional density function of* $Y_2 \mid Y_1 = y_1$ *is obtained by*

$$P(Y_2 = y_2 \mid Y_1 = y_1) = \binom{y_1+y_2}{y_2} \left(\frac{\phi_2}{\phi_1+\phi_2+v(1+\lambda_t)}\right)^{y_2}$$

$$\times \left(\frac{\phi_1+v(1+\lambda_t)}{\phi_1+\phi_2+v(1+\lambda_t)}\right)^{y_1+1},$$

which is the pmf of the NB distribution with parameters $(y_1 + 1)$ *and* $\dfrac{\phi_1+v(1+\lambda_t)}{\phi_1+\phi_2+v(1+\lambda_t)}$. *In the same way, the conditional distribution of* $Y_1 \mid Y_2 = y_2$ *is* $NB\left(y_2+1, \dfrac{\phi_2+v(1+\lambda_t)}{\phi_1+\phi_2+v(1+\lambda_t)}\right)$.

d) *The pgf of* $Z = Y_1 + Y_2$ *can be obtained as:*

$$G_Z(s) = \frac{v(1+\lambda_t)}{v(1+\lambda_t)+(\phi_1+\phi_2)(1-s)}$$

$$= \frac{\dfrac{v(1+\lambda_t)}{\phi_1+\phi_2+v(1+\lambda_t)}}{1 - \dfrac{\phi_1+\phi_2}{\phi_1+\phi_2+v(1+\lambda_t)}s}.$$

Hence, Z has the geometric distribution with parameter $\dfrac{v(1+\lambda_t)}{\phi_1+\phi_2+v(1+\lambda_t)}$.

e) *The distribution of* $Y_i \mid Y_1+Y_2$ *is binomial with parameters* Y_1+Y_2 *and* $\dfrac{\phi_i}{\phi_1+\phi_2}$, *for* $i = 1,2.$

The proof is omitted because it is straight-forward.

Proposition 3. *Suppose that* $\mathbf{Y} \sim BPWE(v,\lambda_t,\phi_1,\phi_2)$, *then*

$$E(Y_i) = \frac{\phi_i}{v(1+\lambda_t)}, \qquad i = 1,2, \qquad (6.3)$$

$$E(Y_i^2) = \frac{\phi_i}{v(1+\lambda_t)}\left(1+\frac{\phi_i}{v(1+\lambda_t)}\right), \qquad i = 1,2, \qquad (6.4)$$

$$Var(Y_i) \quad = \quad \frac{\phi_i(\phi_i + v(1 + \lambda_t))}{v^2(1 + \lambda_t)^2}, \qquad i = 1, 2, \qquad (6.5)$$

$$E(Y_1 Y_2) \quad = \quad \frac{2\phi_1 \phi_2}{v^2(1 + \lambda_t)^2} \qquad\qquad\qquad (6.6)$$

and $Cov(Y_1, Y_2) = \dfrac{\phi_1 \phi_2}{v^2(1 + \lambda_t)^2}$

which is always positive and so the model possesses only non-negative covariances.

Proof. The proof is simply obtained from Proposition 2.   □

However, as we mentioned earlier, the basic bivariate PWE distribution has positive covariance and is appropriate for only bivariate models with positive correlation between two components. Following Gomez Deniz et al. [2012], we derive the Sarmanov bivariate PWE distribution with PWE marginal distribution. In fact, Sarmanov [1966] described a family of bivariate densities. The properties of the Sarmanov family is discussed in Lee [1996]. Lee [1996] showed that the correlation coefficient of this family has wider range than Farlie-Gumbel-Morgenstern (Farlie [1960]).

Now, in the following, we define the Sarmanov bivariate PWE distribution. Suppose that $f_1(y_1)$ and $f_2(y_2)$ are univariate pdf or pmf with supports defined on $\chi_1 \subseteq \mathbb{R}$, $\chi_2 \subseteq \mathbb{R}$, respectively. Also, let $h_i(t)$, $i = 1, 2$ be bounded non-constant functions such that $\int_{-\infty}^{\infty} h_i(t) f_i(t) dt = 0$. Then, the Sarmanov bivariate joint density (or pmf) with given marginal $f_1(y_1)$ and $f_2(y_2)$ is defined by

$$f(y_1, y_2) = f_1(y_1) f_2(y_2) \left\{ 1 + w h_1(y_1) h_2(y_2) \right\}, \qquad (6.7)$$

where w is a real number such that $1 + w h_1(y_1) h_2(y_2) \geq 0$ for all y_1 and y_2.

Lee [1996] proposed some general methods to construct mixing functions $h_i(y_i)$ for different types of marginals. When $f_i(y_i)$, $i = 1, 2$ are defined on $[0, \infty)$, by the Corollary 2 of Lee [1996], $h_i(y_i)$ can be defined as $h_i(y_i) = e^{-y_i} - L_i(1)$, for $y_i \geq 0$, where $L_i(t) = \int_0^{\infty} e^{-t y_i} f_i(y_i) dy_i$ denotes the Laplace transform of f_i, for $i = 1, 2$. The correlation coefficient of Y_1 and Y_2 if it exists, is given by $\rho = \dfrac{w[-L'_1(1) - L_1(1)\mu_1][-L'_2(1) - L_2(1)\mu_2]}{\sigma_1 \sigma_2}$, where μ_i and σ_i are the means and standard deviations of f_i for $i = 1, 2$, $L'(t)$ represents the first derivative of $L(t)$ and w satisfies

$$L_w \leq w \leq U_w. \qquad (6.8)$$

where $L_w = \dfrac{-1}{\max\{L_1(1)L_2(1), (1 - L_1(1))(1 - L_2(1))\}}$, and

$U_w = \dfrac{1}{\max\{L_1(1)(1 - L_2(1)), L_2(1)(1 - L_1(1))\}}$. Therefore, we can derive the

Sarmanov bivariate PWE with PWE marginal distribution by replacing $f_i(y_i)$ and $h_i(y_i)$ for $i = 1, 2$ in (6.7) by

$$f_i(y_i) \sim PWE(v, \lambda_{ti})$$

$$h_i(y_i) = e^{-y_i} - \frac{v(1+\lambda_{ti})}{1 - e^{-1} + v(1+\lambda_{ti})}.$$

Thus, we can define the Sarmanov bivariate PWE random vector as follows.

The random vector $\mathbf{Y} = (Y_1, Y_2)$ has the Sarmanov bivariate PWE distribution (SPWE) if it has the following joint pmf:

$$p(y_1, y_2) = \frac{v^2(1+\lambda_{t1})(1+\lambda_{t2})}{[1+v(1+\lambda_{t1})]^{y_1+1}[1+v(1+\lambda_{t2})]^{y_2+1}}$$

$$\times \left[1 + w \left(e^{-y_1} - \frac{v(1+\lambda_{t1})}{1 - e^{-1} + v(1+\lambda_{t1})} \right) \left(e^{-y_2} - \frac{v(1+\lambda_{t2})}{1 - e^{-1} + v(1+\lambda_{t2})} \right) \right], \quad (6.9)$$

where $y_1, y_2 = 0, 1, 2, \cdots$, $v, \lambda_{t1}, \lambda_{t2} > 0$ and w is as in Equation (6.8). In the model, the means, second-order moments and variances of Y_i are as in Equations (6.3), (6.4) and (6.5), only λ_t is replaced by λ_{ti} and $\phi_1 = \phi_2 = 1$.

According to Lee [1996], the product moment is given by $E(Y_1 Y_2) = \mu_1 \mu_2 + w v_1 v_2$ where $\mu_i = \dfrac{1}{v(1+\lambda_{ti})}$ is the mean of the $PWE(v, \lambda_{ti})$ and $v_i = $

$$\int t h_i(t) f_i(t) dt = \frac{-(1 - e^{-1}) \left(1 + v(1+\lambda_{ti}) \right)}{\left(1 - e^{-1} + v(1+\lambda_{ti}) \right)^2}, \text{ thus } E(Y_1 Y_2) = \frac{1}{v^2(1+\lambda_{t1})(1+\lambda_{t2})} + $$

$$w \frac{(1 - e^{-1})^2 \left(1 + v(1+\lambda_{t1}) \right) \left(1 + v(1+\lambda_{t2}) \right)}{\left(1 - e^{-1} + v(1+\lambda_{t1}) \right)^2 \left(1 - e^{-1} + v(1+\lambda_{t2}) \right)^2} \quad \text{and} \quad Cov(Y_1, Y_2) = $$

$$w \frac{(1 - e^{-1})^2 \left(1 + v(1+\lambda_{t1}) \right) \left(1 + v(1+\lambda_{t2}) \right)}{\left(1 - e^{-1} + v(1+\lambda_{t1}) \right)^2 \left(1 - e^{-1} + v(1+\lambda_{t2}) \right)^2}. \text{ Therefore, depending on the}$$

sign of ω the Sarmanov bivariate PWE distribution has a correlation of any sign. The random vector $\mathbf{Y} = (Y_1, Y_2)$ with pmf (6.9) is denoted by $\mathbf{Y} \sim SPWE(v, \lambda_{t1}, \lambda_{t2}, \omega)$. Now, we want to estimate the parameters of two kinds of bivariate PWE distributions using maximum likelihood method (ML). To find the ML estimation of the basic bivariate PWE distribution parameters, let (y_{1i}, y_{2i}), $i = 1, ..., n$ be the observations of a random sample from $BPWE(v, \lambda_t, \phi_1, \phi_2)$ in (6.2). The log-likelihood function is :

$$l(\psi) \propto n \ln v + \ln(1 + \lambda_t) + n\bar{y}_1 \ln \phi_1 + n\bar{y}_2 \ln \phi_2$$
$$- n(\bar{y}_1 + \bar{y}_2 + 1) \ln(\phi_1 + \phi_2 + v(1 + \lambda_t)), \qquad (6.10)$$

where $\psi = (v, \lambda_t, \phi_1, \phi_2)$.

The ML estimator of ψ is obtained by maximizing $l(\psi)$ with respect to all parameters using numerical methods. The log-likelihood function of (y_{1i}, y_{2i}), $i = 1, ..., n$, the random sample of the Sarmanov bivariate PWE distribution, is

$$l^* = l(v, \lambda_{t1}, \lambda_{t2}, \omega)$$
$$= 2n \ln v + n \ln(1 + \lambda_{t1}) + n \ln(1 + \lambda_{t2})$$
$$- n(\bar{y}_1 + 1) \ln \left(1 + v(1 + \lambda_{t1}) \right) - n(\bar{y}_2 + 1) \ln \left(1 + v(1 + \lambda_{t2}) \right)$$
$$+ \sum \ln \left[1 + w \left(e^{-y_{1i}} - \frac{v(1 + \lambda_{t1})}{1 - e^{-1} + v(1 + \lambda_{t1})} \right) \right.$$
$$\left. \times \left(e^{-y_{2i}} - \frac{v(1 + \lambda_{t2})}{1 - e^{-1} + v(1 + \lambda_{t2})} \right) \right].$$

The ML estimators of the $v, \lambda_{t1}, \lambda_{t2}$ and ω are obtained by maximizing l^* using numerical methods.

6.1.2 Model Development: Part (B) The Bivariate INAR (1) Processes with Paired Bivariate PWE Innovations

Suppose that $Y_t = (Y_{t,1}, Y_{t,2}); t = 1, 2, \cdots$ is a bivariate integer-valued time series (BINAR) where

$$Y_{t,1} = \rho_1 * Y_{t-1,1} + R_{t,1}$$
$$Y_{t,2} = \rho_2 * Y_{t-1,2} + R_{t,2}; \quad \rho_i \in (0, 1), i = 1, 2. \qquad (6.11)$$

We assumed that the innovation vector $R_t = (R_{t,1}, R_{t,2})$ follows two types of introduced distributions - the basic and the Sarmanov bivariate PWE distributions and $R_{t,i}$ is independent from $Y_{s,i}, i = 1, 2$, for each fixed t and $s < t$. Also, innovations are independent from the counting series in binomial thinning operator "$*$". Based on these distributions, the BINAR (1) process with two kinds of the bivariate PWE distributed innovations are introduced. First, we consider the BINAR (1) in (6.11) with $R_t \sim BPWE(v, \lambda_t, \phi_1, \phi_2)$ which is denoted by BINAR (1)-BPWE. The following properties of the model are obtained based on Pedeli and Karlis [2011].

Proposition 4. *Let* $Y_t = (Y_{t,1}, Y_{t,2}); t = 1, 2, \cdots$ *be the BINAR (1)-BPWE process, then*

a) *The covariance function between two terms of innovations and two terms of the process are*

$$Cov(R_{t,1}, R_{t,2}) = \frac{\phi_1 \phi_2}{(v(1 + \lambda_t))^2},$$

$$Cov(Y_{t,1}, Y_{t,2}) = \frac{cov(R_{t,1}, R_{t,2})}{1 - \rho_1 \rho_2} = \frac{\phi_1 \phi_2}{(1 - \rho_1 \rho_2)(v(1 + \lambda_t))^2},$$

which are positive.

b) *Also for $i = 1, 2$, the conditional mean and conditional variance of the component of the process are calculated as*

$$E(Y_{t,i} | Y_{t-1,i}) = \rho_i Y_{t-1,i} + \frac{\phi_i}{v(1 + \lambda_t)},$$

$$Var(Y_{t,i} | Y_{t-1,i}) = \rho_i (1 - \rho_i) Y_{t-1,i} + \frac{\phi_i (\phi_i + v(1 + \lambda_t))}{(v(1 + \lambda_t))^2},$$

and hence,

c)

$$E(Y_{t,i}) = \frac{\phi_i}{(1 - \rho_i) v(1 + \lambda_t)},$$

$$Var(Y_{t,i}) = \frac{\phi_i(\rho_i + 1) v(1 + \lambda_t) + \phi_i^2}{(1 - \rho_i^2)(v(1 + \lambda_t))^2},$$

$$FI(Y_{t,i}) = 1 + \frac{\phi_i}{(\rho_i + 1) v(1 + \lambda_t)},$$

In the above last equation, each component of the vector Y_t is over-dispersed marginally.

d) *The conditional joint pmf of the process is given by*

$$p(y_t | y_{t-1}) = \sum_{k=0}^{u} \sum_{s=0}^{v} p_1(k) p_2(s) P(R_{t,1} = y_{t,1} - k, R_{t,2} = y_{t,2} - s), \qquad (6.12)$$

where $u = min(y_{t,1}, y_{t-1,1})$, $v = min(y_{t,2}, y_{t-1,2})$, and

$$p_1(k) = \binom{y_{t-1,1}}{k} \rho_1^k (1 - \rho_1)^{y_{t-1,1} - k},$$

$$p_2(s) = \binom{y_{t-1,2}}{s} \rho_2^s (1 - \rho_2)^{y_{t-1,2} - s},$$

and $P(R_{t,1} = y_{t,1} - k, R_{t,2} = y_{t,2} - s)$ is given by Equation (6.2) after replacing y_1 by $y_{t,1} - k$ and y_2 by $y_{t,2} - s$.

Now, suppose that the process Y_t is the BINAR (1) in Equation (6.11) where its innovation process follows from Sarmanov bivariate PWE in Equation (6.9) which is denoted by BINAR (1)-SPWE.

Proposition 5. *Based on Pedeli and Karlis [2011], the properties of the BINAR (1)-SPWE process are obtained as:*

a) *the covariance function between two components of R_t is:* $Cov(R_{t,1}, R_{t,2}) =$

$$w \frac{(1-e^{-1})^2 \left(1+v(1+\lambda_{t1})\right)\left(1+v(1+\lambda_{t2})\right)}{\left(1-e^{-1}+v(1+\lambda_{t1})\right)^2 \left(1-e^{-1}+v(1+\lambda_{t2})\right)^2}.$$

b) *The covariance of two components of the BINAR (1) process with Sarmanov distribution Y_t is obtained as*

$$Cov(Y_{t,1}, Y_{t,2}) = w \frac{(1-e^{-1})^2 \left(1+v(1+\lambda_{t1})\right)\left(1+v(1+\lambda_{t2})\right)}{(1-\rho_1\rho_2)\left(1-e^{-1}+v(1+\lambda_{t1})\right)^2 \left(1-e^{-1}+v(1+\lambda_{t2})\right)^2}.$$

It can be seen that these covariance functions can have both negative and positive sign.
Also, for $i = 1, 2$,

c)

$$E(Y_{t,i}|Y_{t-1,i}) = \rho_i Y_{t-1,i} + \frac{1}{v(1-\rho_i)(1+\lambda_{ti})}$$

$$Var(Y_{t,i}|Y_{t-1,i}) = \rho_i(1-\rho_i)Y_{t-1,i} + \frac{1+v(1+\lambda_{ti})}{(v(1+\lambda_{ti}))^2}.$$

d)

$$E(Y_{t,i}) = \frac{1}{v(1-\rho_i)(1+\lambda_{ti})},$$

$$Var(Y_{t,i}) = \frac{(1+\rho_i)v(1+\lambda_{ti})+1}{(1-\rho_i^2)v^2(1+\lambda_{ti})^2},$$

$$FI(Y_{t,i}) = 1 + \frac{1}{v(1-\rho_i)(1+\lambda_{ti})},$$

where FI shows $Y_{t,i}, i = 1, 2$ are over-dispersed.

e)

$$Cov(Y_{t,1}, Y_{t,2} | Y_{t-1,1}, Y_{t-1,2}) = Cov(R_{t,1}, R_{t,2})$$

$$= w \frac{(1 - e^{-1})^2 (1 + v(1 + \lambda_{t1}))(1 + v(1 + \lambda_{t2}))}{(1 - e^{-1} + v(1 + \lambda_{t1}))^2 (1 - e^{-1} + v(1 + \lambda_{t2}))^2}$$

.

f) The conditional joint pmf of the BINAR (1)-SPWE model is also obtained from Equation (6.12), except that $P(R_{t,1} = R_1, R_{t,2} = R_2)$ is replaced by (6.9).

Remark 1. *As it is mentioned in Mamode Khan et al. [2019], the stationary conditions for these models are $\rho_i < 1$, $i = 1,2$ and under these conditions, it is easily seen that $E(Y_{t,i})$, $i = 1,2$ and $Cov(Y_{t,1}, Y_{t,2})$ do not depend on t and $Var(Y_{t,i}) < \infty$.*

Now, we want to estimate the parameters of the considered BINAR (1) processes by two methods.

6.2 Estimation of the Parameters of the BINAR (1) Processes

In this section, two estimation methods are presented for both considered BINAR (1) models.

6.2.1 CLS and CML Estimations for BINAR (1)-BPWE Process

Suppose that $\{Y_t\}_{t=1}^n$ is a random sample from the *BINAR* $(1) - BPWE$ process. To have an identifiable model, we re-parameterized the model by setting $\mu_i = \frac{\phi_i}{v(1 + \lambda_t)}; i = 1, 2$, and hence the vector of parameters become $\beta = (\rho_1, \rho_2, \mu_1, \mu_2)$. Then, the least squares estimator of β is obtained by minimizing the following quantities with respect to the parameter vector.

$$Q_i = \sum_{t=2}^n [Y_{t,i} - E(Y_{t,i} | Y_{t-1,i})]^2$$

$$= \sum_{t=2}^n [Y_{t,i} - \rho_i Y_{t-1,i} - \mu_i)]^2; i = 1, 2.$$

After some calculations, the CLS estimators are given by

$$\hat{\rho}_i^{CLS} = \frac{(n-1)\sum_{t=2}^n Y_{t,i}Y_{t-1,i} - \sum_{t=2}^n Y_{t,i}\sum_{t=2}^n Y_{t-1,i}}{(n-1)\sum_{t=2}^n Y_{t-1,i}^2 - (\sum_{t=2}^n Y_{t-1,i})^2},$$

$$\hat{\mu}_i^{CLS} = \frac{1}{n-1}(\sum_{t=2}^n Y_{t,i} - \hat{\rho}_i^{CLS}\sum_{t=2}^n Y_{t-1,i}).$$

According to Zhang et al. [2020], we can obtain the asymptotic behavior of the estimators $\hat{\beta}^{CLS} = (\hat{\rho}_1^{CLS}, \hat{\rho}_2^{CLS}, \hat{\mu}_1^{CLS}, \hat{\mu}_2^{CLS})$.

Remark 2. *Under regularity conditions which are given in Klimko and Nelson [1978], $\hat{\beta}^{CLS}$ is consistent estimator and has the following asymptotic distributions:*

$$\sqrt{n}(\hat{\beta}^{CLS} - \beta) \longrightarrow^d N(\mathbf{0}, V^{-1}WV^{-1}), \tag{6.13}$$

where $W = E\left(\frac{\partial g_t^T}{\partial \beta} U_t U_t^T \frac{\partial g_t}{\partial \beta^T}\right)$, $V = E\left(\frac{\partial g_t^T}{\partial \beta} \frac{\partial g_t}{\partial \beta^T}\right)$, $g_t = (g_1, g_2)$, $U_t = (U_1, U_2)$, $g_i = E(Y_{t,i}|Y_{t-1,i}) = \rho_i Y_{t,i} + \psi_i, i = 1, 2$ *and* $U_i = Y_{t,i} - g_i$, $i = 1, 2$.

To obtain the conditional maximum likelihood estimators, first we rewrite the pmf of R_t based on the parameters μ_1 and μ_2 as follows:

$$P(R_{t,1} = r_1, R_{t,2} = r_2) = \frac{(r_1 + r_2)!}{r_1! r_2!} \frac{\mu_1^{r_1}\mu_2^{r_2}}{(\mu_1 + \mu_2 + 1)^{r_1 + r_2 + 1}}, r_1, r_2 = 0, 1, \cdots, \tag{6.14}$$

where $\mu_i = \frac{\phi_i}{v(1+\lambda_t)}; i = 1, 2$.

The conditional likelihood function is given by

$$L(\beta) = \prod_{t=2}^n p(y_t|y_{t-1}), \tag{6.15}$$

where $p(y_t|y_{t-1})$ is calculated by Equation (6.12) with replacing Equation (6.14) in it. The conditional maximum likelihood estimation of β is driven by maximizing Equation (6.15) by numerical methods.

6.2.2   CLS and CML Estimations for BINAR (1)-SPWE Process

Let $\{Y_t\}_{t=1}^n$ be a random sample of BINAR (1)-SPWE process, then the CLS estimator of $\beta^* = (\rho_1, \rho_2, \tau_1, \tau_2, w)$, where $\tau_i = v(1 + \lambda_{t i})$, is driven by minimizing Q_1, Q_2 and Q_3 where for $i = 1, 2$

$$\begin{aligned} Q_i &= \sum_{t=2}^n [Y_{t,i} - E(Y_{t,i}|Y_{t-1,i})]^2, \\ &= \sum_{t=2}^n [Y_{t,i} - \rho_i Y_{t-1,i} - \frac{1}{\tau_i}]^2. \end{aligned}$$

and

$$Q_3 = \sum_{t=2}^{n} \{[Y_{t,1} - E(Y_{t,1}|Y_{t-1,1})][Y_{t,2} - E(Y_{t,2}|Y_{t-1,2})]$$
$$- Cov(Y_{t,1}, Y_{t,2}|Y_{t-1,1}, Y_{t-1,2})\}^2$$
$$= \sum_{t=2}^{n} \{[Y_{t,1} - \rho_1 Y_{t-1,1} - \frac{1}{\tau_1}][Y_{t,2} - \rho_2 Y_{t-1,2} - \frac{1}{\tau_2}]$$
$$- w\frac{(1-e^{-1})^2(1+\tau_1)(1+\tau_2)}{(1-e^{-1}+\tau_1)^2(1-e^{-1}+\tau_2)^2}\}^2.$$

After some calculation, the estimators are derived as follows:

$$\hat{\rho}_i^{CLS} = \frac{(n-1)\sum_{t=2}^{n} Y_{t,i} Y_{t-1,i} - \sum_{t=2}^{n} Y_{t,i}\sum_{t=2}^{n} Y_{t-1,i}}{(n-1)\sum_{t=2}^{n} Y_{t-1,i}^2 - (\sum_{t=2}^{n} Y_{t-1,i})^2},$$

$$\hat{\tau}_i^{CLS} = \left(\frac{1}{n-1}\left[\sum_{t=2}^{n} Y_{t,i} - \hat{\rho}_i^{CLS}\sum_{t=2}^{n} Y_{t-1,i}\right]\right)^{-1}.$$

and

$$\hat{w}^{CLS} = \frac{1}{n-1}\sum_{t=2}^{n}\{[Y_{t,1} - \hat{\rho}_1^{CLS} Y_{t-1,1} - \frac{1}{\hat{\tau}_1^{CLS}}][Y_{t,2} - \hat{\rho}_2^{CLS} Y_{t-1,2} - \frac{1}{\hat{\tau}_2^{CLS}}]$$
$$\times \frac{(1-e^{-1}+\hat{\tau}_1^{CLS})^2(1-e^{-1}+\hat{\tau}_2^{CLS})^2}{(1-e^{-1})^2(1+\hat{\tau}_1^{CLS})(1+\hat{\tau}_2^{CLS})}.$$

Similar to BINAR (1)-BPWE, we can derive the asymptotic behavior of the estimators $\hat{\beta}_*^{CLS}$.

Remark 3. *Under regularity conditions in Klimko and Nelson [1978], $\hat{\beta}_*^{CLS}$ is consistent estimator and:*

$$\sqrt{n}(\hat{\beta}_*^{CLS} - \beta^*) \longrightarrow^d N(\mathbf{0}, V^{*-1}WV^{*-1}), \tag{6.16}$$

where $W^* = E(\frac{\partial g_t^{*T}}{\partial \beta^*} U_t^* U_t^{*T} \frac{\partial g_t^*}{\partial \beta^{*T}})$, $V^* = E(\frac{\partial g_t^{*T}}{\partial \beta^*} \frac{\partial g_t^*}{\partial \beta^{*T}})$, $g_t^* = (g_1^*, g_2^*, g_w^*)$, $U_t^* = (U_1^*, U_2^*, U_w^*)$, $g_i^* = E(Y_{t,i}|Y_{t-1,i}) = \alpha_i Y_{t,i} + \frac{1}{\tau_i}, i = 1, 2, g_w^* = Cov(Y_{t,1}, Y_{t,2}|Y_{t-1,1}, Y_{t-1,2}) = \frac{w(1-e^{-1})^2(1+\tau_1)(1+\tau_2)}{(1-e^{-1}+\tau_1)^2(1-e^{-1}+\tau_2)^2}$. *Also,* $U_i^* = Y_{t,i} - g_i^*$, $i = 1, 2$, *and* $U_w^* = U_1^* U_2^* - g_i^*, i = 1, 2$.

To derive the CMLEs of the parameters by applying the new parameters $\tau_i = v(1+\lambda_{ti}), i = 1, 2$, the pmf of R_t changes into

$$P(R_{t,1} = r_1, R_{t,1} = r_2) = \frac{\tau_1\tau_2}{(1+\tau_1)^{r_1+1}(1+\tau_2)^{r_2+1}} \tag{6.17}$$
$$\times \left[1 + w(e^{-r_1} - \frac{\tau_1}{1-e^{-1}+\tau_1})(e^{-r_2} - \frac{\tau_2}{1-e^{-1}+\tau_2})\right].$$

Now, suppose that $\{Y_t\}_{t=1}^n$ is a random sample of BINAR (1)-SPWE process, then the conditional likelihood function is

$$L(\beta^*) = \prod_{t=2}^n p(y_t|y_{t-1}),$$ (6.18)

where $p(y_t|y_{t-1})$ is given by Equation (6.12) when the join pmf of R_t in Equation (6.12) is replaced by Equation (6.17). By maximizing Equation (6.18) with respect to β^*, we can obtain the CML estimation of the model parameters. This can be achieved by numerical methods.

6.3 Simulation Study

In order to compare the performance of the two estimation methods, CLS and CML, for two bivariate INAR (1) models proposed in the previous section, we first randomly simulate $Y_1, Y_2, \cdots Y_n$ from the models and then calculate the CLS and CML estimators. The simulation study is based on the number of replication $B = 10000$ and sample sizes $n = 100, 300, 500$. The true values of parameters $\beta = (\rho_1, \rho_2, \mu_1, \mu_2)$ and $\beta^* = (\rho_1, \rho_2, \tau_1, \tau_2, w)$ for two models are considered as follows.

For BINAR (1)-BPWE:
   S1: $(0.1, 0.3, 0.5, 1.25)$, S2:$(0.3, 0.5, 0.5, 1.25)$, S3: $(0.1, 0.3, 1, 2)$,
   S4:$(0.3, 0.5, 1, 2)$,
For BINAR (1)-SPWE:
   S5: $(0.1, 0.3, 0.3, 1, -0.5)$, S6: $(0.3, 0.1, 0.3, 1, -0.5)$, S7: $(0.1, 0.3, 1, 2, 0.5)$,
and
   S8: $(0.3, 0.5, 1, 2, 0.5)$.

For each model, the CML and CLS estimates, bias and mean square error (MSE) of the estimators are calculated. The results for BINAR (1)-BPWE and BINAR (1)-SPWE are given in the Tables 6.1 and 6.2, respectively. The results show that the MSE of estimators decrease as sample size increases which indicates the consistency of the estimators. Moreover, for the two models, the MSE of the CML estimators of all parameters are smaller than MSE of the CLS estimators (except for few cases) which shows that the CML estimator has better performance.

Model		n=100 CML Bias	n=100 CML MSE	n=100 CLS Bias	n=100 CLS MSE	n=300 CML Bias	n=300 CML MSE	n=300 CLS Bias	n=300 CLS MSE	n=500 CML Bias	n=500 CML MSE	n=500 CLS Bias	n=500 CLS MSE
S1	ρ_1	−0.00296	0.00789	−0.01351	0.01137	−0.00337	0.00351	−0.00452	0.00387	−0.00329	0.00225	−0.00342	0.00233
	ρ_2	−0.02261	0.01039	−0.01988	0.01047	−0.00848	0.00356	−0.00746	0.00356	−0.00477	0.00217	−0.00421	0.00216
	μ_1	0.00742	0.01078	0.00746	0.01093	0.00215	0.00354	0.00213	0.00356	0.00156	0.00215	0.00155	0.00216
	μ_2	0.03382	0.05931	0.03415	0.06024	0.01208	0.01971	0.01211	0.01979	0.00728	0.01167	0.00737	0.01170
S2	ρ_1	−0.02557	0.01234	−0.02274	0.01256	−0.00920	0.00420	−0.00820	0.00420	−0.00514	0.00253	−0.00457	0.00254
	ρ_2	−0.03174	0.00964	−0.02676	0.00945	−0.01008	0.00305	−0.00844	0.00304	−0.00611	0.00185	−0.00511	0.00185
	μ_1	0.01283	0.01273	0.01304	0.01286	0.00328	0.00418	0.00330	0.00420	0.00291	0.00255	0.00295	0.00256
	μ_2	0.06158	0.08319	0.06185	0.08436	0.01873	0.02645	0.01884	0.02659	0.01118	0.01597	0.01120	0.01601
S3	ρ_1	−0.00490	0.00738	−0.01477	0.01061	−0.00379	0.00323	−0.00478	0.00356	−0.00263	0.00214	−0.00270	0.00221
	ρ_2	−0.02262	0.00989	−0.01964	0.00999	−0.00808	0.00336	−0.00706	0.00337	−0.00403	0.00205	−0.00344	0.00205
	μ_1	0.01317	0.03226	0.013400	0.03282	0.00531	0.01074	0.00534	0.01077	0.00233	0.00664	0.00233	0.00666
	μ_2	0.04795	0.01380	0.04830	0.01400	0.01972	0.04661	0.01960	0.04683	0.00826	0.02761	0.00828	0.02767
S4	ρ_1	−0.02247	0.01075	−0.01957	0.01089	−0.00745	0.00367	−0.00650	0.00368	−0.00408	0.00221	−0.00350	0.00221
	ρ_2	−0.03093	0.00910	−0.02588	0.00890	−0.00990	0.00289	−0.00823	0.00288	−0.00604	0.00174	−0.00502	0.00174
	μ_1	0.02303	0.04114	0.02299	0.04162	0.00804	0.01347	0.00797	0.01352	0.00467	0.00802	0.00472	0.00804
	μ_2	0.09365	0.19687	0.093503	0.01993	0.03020	0.06276	0.03007	0.06316	0.02033	0.03785	0.02036	0.03795

TABLE 6.1
Bias and MSE of the estimators for BINAR (1)-BPWE

Model		n=100				n=300				n=500			
		CML		CLS		CML		CLS		CML		CLS	
		Bias	MSE	Bias	MSE	Bias	MSE	Bias	MSE	Bias	MSE	Bias	MSE
S5	ρ_1	−0.00343	0.00694	−0.01350	0.01030	−0.00394	0.00299	−0.00487	0.00331	−0.00271	0.00199	−0.00275	0.00205
	ρ_2	−0.02446	0.01098	−0.02168	0.01111	−0.00832	0.003722	−0.00734	0.00372	−0.00477	0.00220	−0.00416	0.00221
	τ_1	0.07438	0.00865	0.07442	0.00870	0.07303	0.00632	0.07304	0.00633	0.07287	0.00592	0.07286	0.00592
	τ_2	0.01420	0.04209	0.01488	0.04304	0.00312	0.01318	0.00314	0.01324	0.00390	0.00804	0.00397	0.00807
	w	0.18598	0.50673	0.04230	0.93877	0.09355	0.26490	0.04651	0.31419	0.06293	0.18258	0.05239	0.19206
S6	ρ_1	−0.02139	0.00981	−0.01862	0.00993	−0.00751	0.00326	−0.00651	0.00327	−0.00377	0.00192	−0.00318	0.00193
	ρ_2	−0.03240	0.00995	−0.02727	0.00977	−0.01000	0.00307	−0.00827	0.00305	−0.00667	0.00187	−0.0056	0.00186
	τ_1	0.07370	0.00948	0.07377	0.00956	0.07308	0.00666	0.07310	0.00667	0.07289	0.00611	0.07289	0.00612
	τ_2	0.00268	0.05355	0.00323	0.05445	0.00174	0.01746	0.001896	0.01755	0.00135	0.00994	0.00137	0.00996
	w	0.24724	0.64873	0.06997	1.10719	0.12980	0.39264	0.05825	0.36405	0.06877	0.28288	0.056855	0.22335
S7	ρ_1	−0.00495	0.00717	−0.01468	0.01034	−0.00374	0.00332	−0.00469	0.00364	−0.00289	0.00216	−0.00291	0.00222
	ρ_2	−0.02572	0.01251	−0.02303	0.01271	−0.00896	0.00424	−0.00798	0.00425	−0.00453	0.002499	−0.00393	0.00250
	τ_1	0.02173	0.03549	0.02201	0.03620	0.01181	0.01134	0.01193	0.01141	0.00795	0.00675	0.00792	0.00677
	τ_2	0.051206	0.23033	0.05228	0.23448	0.01524	0.07038	0.01532	0.07085	0.01369	0.04180	0.01372	0.04191
	w	−0.21141	0.53344	−0.01523	1.17530	−0.07539	0.27203	−0.00069	0.37959	−0.04135	0.18836	−0.0064	0.23015
S8	ρ_1	−0.02185	0.01075	−0.01892	0.01086	−0.00739	0.00372	−0.00643	0.00372	−0.00450	0.00217	−0.00391	0.00218
	ρ_2	−0.03376	0.01113	−0.02866	0.01094	−0.01198	0.00358	−0.01023	0.00355	−0.00719	0.00217	−0.00618	0.00216
	τ_1	0.02121	0.04292	0.02175	0.04361	0.01025	0.01378	0.01022	0.01385	0.00910	0.00799	0.00909	0.00802
	τ_2	0.03197	0.26713	0.03391	0.27178	0.00884	0.08498	0.00912	0.08539	0.00362	0.04992	0.00359	0.04999
	w	−0.23785	0.66562	0.02491	1.37989	−0.10836	0.40217	−0.00108	0.46900	−0.04032	0.29032	0.00013	0.28002

TABLE 6.2
Bias and MSE of the estimators for BINAR (1)-SPWE

6.4 Application to Daily Stock Transactions Series from New York Stock Exchange

In this section, we consider the first 150 count observations from the AXP and GE intra-day stock transaction series from the August 1995 New York Stock Exchange (NYSE) used by Nytholm [2003] and available in the link $http : //qed.econ.queensu.ca/jae/2003 - v18.4/nyholm/$. The means, variances and FIs of AXP-NUM and GE-NUM of transactions are in Table 6.3. The results show that the FIs are greater than 1 and then the two time series data seem to be over-dispersed. To check this, we applied the over-dispersion test introduced by Schweer and Weiß [2014] and obtained the p-values equal to 7.8304×10^{-8} and 0.0000, respectively, which confirmed the over-dispersion of two time series.

Now, we compare the goodness of fit criteria of two proposed BINAR (1) models with three other BINAR (1) models based on binomial thinning operators with bivariate Poisson (BINAR (1)-BP), bivariate Negative Binomial distributed innovations (BINAR (1)-BNB) [Pedeli and Karlis, 2011], and bivariate Poisson-Lindley distributed innovations (BINAR (1)-PL) (Mamode Khan et al. [2019]) for both time series.

The CML estimates, logarithm of likelihood function (log-lik), AIC, BIC and the RMSE of differences of observations and predicted values of each model are reported in the Table 6.4. The CML estimates are obtained by using function *"nlminb"* in R with initial values obtained through CLS method.

The results show that the BINAR (1)-SPWE has the largest log-lik and the smallest AIC and BIC values. Hence we can conclude that this model has the best fit to the data.

To check the adequacy of the BINAR (1)-SPWE model, we calculate the Pearson residuals

$$e_{t,i} = \frac{Y_{t,i} - \hat{E}(Y_{t,i}|Y_{t-1,i})}{\sqrt{\hat{Var}(Y_{t,i}|Y_{t-1,i})}}, i = 1, 2, \quad t = 2, ..., 150, \tag{6.19}$$

Since the ACFs of two residuals lie between bounds, exactly and the cumulative periodograms are between two confidence bounds, by Zhang et al. [2020], it can be seen that the two sequences of Pearson residuals are white-noise processes. Thus, the BINAR (1)-SPWE model is an appropriate model for these data sets.

Time Series	Mean	Variance	FI
AXP-NUM	2.8333	4.9317	1.7406
GE-NUM	7.6866	22.4716	2.9234

TABLE 6.3
Mean, Variance and FI of two time seies data

Model	CML	Log-lik	AIC	BIC	RMSE	
					AXP-NUM	GE-NUM
BINAR (1)-BP	$\hat{\mu}_{t_1}$ =1.6855 $\hat{\mu}_{t_2}$ =4.9556 $\hat{\phi}$ =0.6775 $\hat{\rho}_1$ =0.3764 $\hat{\rho}_2$ =0.3024	-788.156	1586.31	1586.3323	1.9899	4.359
BINAR (1)-BNB	$\hat{\mu}_{t_1}$ =1.5400 $\hat{\mu}_{t_2}$ =4.4476 $\hat{\beta}$ =0.4338 $\hat{\rho}_1$ =0.3897 $\hat{\rho}_2$ =0.3433	-729.988	1469.996	1470.0174	3.3918	4.3632
BINAR (1)-BPL	$\hat{\lambda}_{t_1}$ =0.9548 $\hat{\lambda}_{t_2}$ =0.3503 $\hat{\rho}_{12}$ =-0.9291 $\hat{\rho}_1$ =0.4292 $\hat{\rho}_2$ =0.3397	-726.86	1463.72	1463.7415	1.9827	4.3194
BINAR (1)-BPWE	$\hat{\mu}_1$ =1.5340 $\hat{\mu}_2$ =1.4173 $\hat{\rho}_1$ =0.4306 $\hat{\rho}_2$ =0.3914	-739.292	1486.584	1488.6054	1.9845	4.3121
BINAR (1)-SPWE	$\hat{\tau}_1$ =0.6445 $\hat{\tau}_2$ =0.2231 $\hat{\omega}_1$ =1.0000 $\hat{\rho}_1$=0.4256 $\hat{\rho}_2$ =0.4128	-723.331	1456.668	1456.6895	1.9845	4.2991

TABLE 6.4
CML estimates, log-lik, AIC, BIC, and RMSE for the models

6.5 Concluding Remarks

In this chapter, first we introduced BPWE and SPWE distributions and studied some of their statistical properties. Then, we proposed two BINAR (1) models with BPWE and SPWE distributed innovations. The parameters were estimated using CML and CLS methods. The performance of the mentioned methods were evaluated by some simulation studies. The results showed that the CML estimator had smaller MSE and hence is better than the CLS estimators. Finally, using a real data application, we showed that the BINAR (1)-SPWE had the best goodness of fit criteria among the five considered BINAR (1) models .

7

Bivariate Poisson Generalized Lindley Distribution and the Associated BINAR (1) Process

Modeling probabilistic behavior of two random variables simultaneously demanded the construction of bivariate random variables. Bivariate distributions may be thus considered as extensions of univariate distributions. Many approaches have been discussed in the literature for the construction of bivariate random variables. Balakrishna and Lai [2009] mentions many of those. Constructing discrete as well as continuous bivariate random variables using mixture methodology is discussed frequently in the statistical literature (see Karlis and Xekalaki [2005], Lai [2006], Sarabia Alegría et al. [2008], etc). The main advantage of this method is that it will have simple expressions for its marginal probability density function (pdf) and hence its moments, correlation, etc. Another method is to use a new class of distribution for the construction. The Sarmanov family (Sarmanov [1966]) of distributions can be used to construct discrete as well as continuous bivariate distributions with a flexible covariance structure. Ting Lee [1996] studied various properties regarding the family and constructed bivariate distributions by using various marginal distributions and construction methods. Also the Farlie-Gumbel-Morgenstern (FGM) copula is a special case of Sarmanov family. The INAR (1) model is suitable for time series count datasets, which exhibit over-dispersion, in many statistical and applied fields such as medical insurance, sports, finance, etc. Since the initial works of McKenzie [1985] and Al Osh and Alzaid [1987] on INAR (1) process with Poisson innovations, significant number of works has been introduced in literature having univariate innovation distributions (see, Livio et al. [2018], Altun [2020], Eliwa et al. [2020], Emrah and Mamode Khan [2021], Irshad et al. [2021a], etc).

Irshad et al. [2021b] proposed a discrete univariate distribution, the two parameter Poisson generalized Lindley (TPPGLD) by mixing Poisson distribution with a new generalized Lindley distribution of Abouammoh et al.'s [2015]. They introduced a count regression model and an INAR (1) model based on TPPGLD (INAR (1) TPPGL) for modeling over dispersed datasets. Also it is proved that INAR (1) TPPGL provides better model than many other recently developed INAR (1) models based on some model selection criterion which immensely motivates us to propose a BINAR (1) process with bivariate TPPGL innovations. Hence in this book, discrete bivariate distributions based on TPPGLD are constructed by using the mixture methodology (the basic bivariate Poisson generalized Lindley (BPGL)) as well as

DOI: 10.1201/9781003677451-7

using the Sarmanov family of distributions (the Sarmanov bivariate Poisson general-ized Lindley (SPGL)). Most importantly, these established distributions are mounted as innovations for BINAR (1) (that is., BINAR (1) BPGL and BINAR (1) SPGL) and both the processes are compared with some other recently proposed BINAR (1) models as well as discussed in Pedeli and Karlis [2011].

We first define the development of TPPGLD and associated INAR (1) process. Then bivariate versions are constructed and adapt it to BINAR (1) with bivariate PGL innovations (BPGL and SPGL) by inducing a cross correlation between the counting series by assuming the paired PGL innovations are jointly distributed.

7.1   Development of the TPPGLD and Associated INAR (1) Process

A generalized Lindley distribution (GLD) with parameters v and λ_t introduced by Abouammoh et al.'s [2015] have pdf,

$$f(y) = \frac{\lambda_t^v y^{v-2}}{(\lambda_t + 1)\Gamma(v)}(y + v - 1)e^{-\lambda_t y}, \quad \lambda_t \geq 0, v \geq 1, y \geq 0. \qquad (7.1)$$

Recently, Irshad et al. [2021b] introduced a mixture distribution by mixing Poisson with GLD, that is, let Y denote the random variable having the TPPGL distribution such that,

$$Y|\lambda \sim P(\lambda)$$

and

$$\lambda|v, \lambda_t \sim GLD(v, \lambda_t)$$

where $\lambda > 0, \lambda_t \geq 0, v \geq 1$. Then the unconditional probability mass function (pmf) of Y having TPPGLD is

$$P(Y = y) = \frac{\lambda_t^v \Gamma(y + v - 1)}{\Gamma(v)y!(\lambda_t + 1)^{y+v+1}}(y + (v - 1)(\lambda_t + 2)), y = 0, 1, 2, \dots \qquad (7.2)$$

Thus the probability generating function (pgf), moment generating function (mgf), and hence the mean and variance of Y are obtained such that,

$$G_y(s) = \frac{\lambda_t^v}{1 + \lambda_t} \frac{2 - s + \lambda_t}{(1 - s + \lambda_t)^v},$$

$$M_Y(t) = \frac{\lambda_t^v}{1 + \lambda_t} \frac{2 - e^{\lambda_t} + \lambda_t}{(1 - e^{\lambda_t} + \lambda_t)^v}, \qquad (7.3)$$

$$E(Y) = \frac{1 + (v - 1)(\lambda_t + 1)}{\lambda_t(1 + \lambda_t)}, \qquad (7.4)$$

and

$$\text{Var}(Y) = \frac{v(1+\lambda_t)}{\lambda_t^2} - \frac{2+\lambda_t}{(1+\lambda_t)^2}. \tag{7.5}$$

Also, Fisher index of dispersion of Y is,

$$\text{DI}(Y) = 1 + \frac{1}{\lambda_t} + \frac{1}{(1+\lambda_t)(v+(v-1)\lambda_t)}, \tag{7.6}$$

which indicates clear case of over-dispersion (since DI>1). Therefore TPPGLD is used as an innovation distribution for a first-order integer-valued autoregressive process (INAR (1) TPPGL) and it is proved that INAR (1) TPPGL gives better results for AIC, empirical mean and variance, than many other recently proposed INAR (1) processes based on real count data sets. The process $\{Y_t\}_{t \in \mathbb{Z}}$ which follows,

$$Y_t = \rho * Y_{t-1} + R_t, 0 \le \rho < 1, \tag{7.7}$$

where R_t is an iid sequence of random variables having TPPGL distribution is said to be INAR (1) TPPGL process. Also R_t is independent of Y_{t-k} for all $k \ge 1$ and counting series in the binomial thinning operator $*$. Now the one step transition probability of INAR (1) TPPGL process is,

$$P(Y_t = k \mid Y_{t-1} = l) = \sum_{i=1}^{\min(k,l)} \binom{l}{i} \rho^i (1-\rho)^{l-i}$$
$$\frac{\lambda_t^{v} \Gamma((k-i)+v-1)((k-i)+(v-1)(\lambda_t+2))}{\Gamma(v)(k-i)!(\lambda_t+1)^{(k-i)+v+1}}, \tag{7.8}$$

The mean, variance, Fisher index of dispersion, conditional mean and conditional variance are given by :

$$E(Y_t) = \frac{1+(v-1)(\lambda_t+1)}{\lambda_t(1+\lambda_t)(1-\rho)}, \tag{7.9}$$

$$\text{Var}(Y_t) = \frac{\rho(1+(v-1)(\lambda_t+1))}{(1-\rho^2)\lambda_t(1+\lambda_t)} + \frac{v(1+\lambda_t)}{(1-\rho^2)\lambda_t^2} - \frac{2+\lambda_t}{(1-\rho^2)(1+\lambda_t)^2}. \tag{7.10}$$

$$\text{DI}(Y_t) = 1 + \frac{1}{\lambda_t(1+\rho)} + \frac{1}{(1+\rho)(1+\lambda_t)(v+(v-1)\lambda_t)}, \tag{7.11}$$

$$E(Y_t \mid Y_{t-1}) = \rho Y_{t-1} + \frac{1+(v-1)(\lambda_t+1)}{\lambda_t(1+\lambda_t)}, \tag{7.12}$$

and

$$\text{Var}(Y_t \mid Y_{t-1}) = \rho(1-\rho)Y_{t-1} + \frac{v(1+\lambda_t)}{\lambda_t^2} - \frac{2+\lambda_t}{(1+\lambda_t)^2}. \tag{7.13}$$

Here also DI greater than 1 implying over-dispersion of the process.

7.2 The Basic Bivariate Poisson Generalized Lindley Distribution and Corresponding BINAR (1) Process

The basic bivariate PGL (BPGL) distribution based on TPPGL is developed in this section.

Proposition 1. *Suppose* $Y = (Y_1, Y_2)$ *denotes a bivariate random vector that possesses basic bivariate PGL (BPGL) distribution such that,*

$$Y_i \mid \Lambda = \lambda \sim P(\lambda \phi_i), i = 1, 2, \text{ independent}$$

and

$$\Lambda \sim GLD(v, \lambda_t),$$

$\phi_i > 0, \lambda_t \geq 0, v \geq 1$. *Then the unconditional pmf of Y having BPGL distribution is*

$$P(Y_1 = y_1, Y_2 = y_2)$$
$$= \frac{\lambda_t{}^v \phi_1^{y_1} \phi_2^{y_2} \Gamma(v + y_1 + y_2 - 1)((v-1)(\lambda_t + \phi_1 + \phi_2 + 1) + y_1 + y_2)}{(\lambda_t + \phi_1 + \phi_2)^{v+y_1+y_2}(\lambda_t + 1)y_1!y_2!\Gamma(v)}$$

(7.14)

where $y_1, y_2 = 0, 1, 2, ..., \phi_1, \phi_2, \lambda_t > 0$ *and* $v \geq 1$.

Proof. By the procedure defined in Gomez-Deniz et al. [2012] the pmf of BPGL is,

$$P(Y_1 = y_1, Y_2 = y_2) = \int_0^\infty P(Y_1 = y_1, Y_2 = y_2 \mid \Lambda = \lambda) dF_\lambda(\lambda)$$

$$= \int_0^\infty \frac{(\lambda \phi_1)^{y_1} e^{-\lambda \phi_1}}{y_1!} \frac{(\lambda \phi_2)^{y_2} e^{-\lambda \phi_2}}{y_2!} \frac{\lambda_t{}^v y^{v-2}}{(\lambda_t + 1)\Gamma(v)}(\lambda + v - 1)e^{-\lambda_t \lambda} d\lambda \quad (7.15)$$

$$= \frac{\lambda_t{}^v \phi_1^{y_1} \phi_2^{y_2} \Gamma(v + y_1 + y_2 - 1)((v-1)(\lambda_t + \phi_1 + \phi_2 + 1) + y_1 + y_2)}{(\lambda_t + \phi_1 + \phi_2)^{v+y_1+y_2}(\lambda_t + 1)y_1!y_2!\Gamma(v)}$$

Hence it is proved.   □

The random vector Y with pmf (7.14) is hereafter denoted by $Y \sim$ BPGL$(\lambda_t, v, \phi_1, \phi_2)$.

Proposition 2. *Suppose* $Y \sim BPGL(\lambda_t, v, \phi_1, \phi_2)$, *then*

1. The marginal pmf of Y_i,

$$P(Y_i = y_i) = \frac{\lambda_t{}^v \phi_i^{y_i} \Gamma(v + y_i - 1)((v-1)(\lambda_t + \phi_i + 1) + y_i)}{(\lambda_t + 1)(\lambda_t + \phi_i)^{v+y_i}\Gamma(v)y_i!},$$

(7.16)

$$y_i = 0, 1, 2, ...$$

Remark 1. *If $\phi_i = 1$ for $i = 1, 2$, $Y_i \sim TPPGL(v, \lambda_t)$.*

2. *The conditional pmf of $Y_2 | Y_1$ for given $y_1 = 0, 1, 2, ...,$*

$$
P(Y_2 = y_2 | Y_1 = y_1) = \frac{\Gamma(y_1 + y_2 + v - 1)\phi_2^{y_2}(\lambda_t + \phi_1)^{v + y_1}}{\Gamma(y_1 + v - 1)\Gamma(y_2 + 1)(\lambda_t + \phi_1 + \phi_2)^{(v + y_1 + y_2)}}
$$
$$
\times \frac{(y_1 + y_2 + (v - 1)(1 + \lambda_t + \phi_1 + \phi_2))}{(y_1 + (v - 1)(1 + \lambda_t + \phi_1))}, y_2 > 0
$$

(7.17)

Similarly, the pmf of $Y_1 | Y_2$ can be obtained.

3. *The joint pgf of $\mathbf{Y}$ is*

$$
G_Y(s_1, s_2) = E(s_1^{Y_1} s_2^{Y_2})
$$

(7.18)

$$
= \frac{\lambda_t^v(\lambda_t - (s_1 - 1)\phi_1 - (s_2 - 1)\phi_2 + 1)}{(\lambda_t + 1)(\lambda_t - (s_1 - 1)\phi_1 - (s_2 - 1)\phi_2)^v}.
$$

(7.19)

Following Proposition 2, we can find the moment related properties as follows, The mean of Y_i,

$$
E(Y_i) = \frac{(v\lambda_t + v - \lambda_t)\phi_i}{\lambda_t(\lambda_t + 1)}, i = 1, 2.
$$

(7.20)

and

$$
E(Y_i^2) = \frac{\phi_i \left[(v - 1)\lambda_t^2 + v\lambda_t + v\phi_i((v - 1)\lambda_t + v + 1) \right]}{\lambda_t^2(\lambda_t + 1)} i = 1, 2.
$$

(7.21)

Hence, the variance of Y_i,

$$
\text{Var}(Y_i) = \phi_i \left(\frac{v}{\lambda_t^2(\lambda_t + 1)} - \frac{1}{\lambda_t(\lambda_t + 1)^2} \right) + \phi_i^2 \left(\frac{v}{\lambda_t^2} - \frac{1}{(\lambda_t + 1)^2} \right) i = 1, 2.
$$

(7.22)

Covariance of Y_1 and Y_2,

$$
\text{Cov}(Y_1, Y_2) = \frac{\phi_1 \phi_2 \left(v(\lambda_t + 1)^2 - \lambda_t^2 \right)}{\lambda_t^2(\lambda_t + 1)^2},
$$

(7.23)

which is always positive.

7.2.1 Estimation of the Parameters of BPGL Distribution

The method of maximum likelihood (MLE) is used to estimate the unknown parameters. Suppose (Y_{1i}, Y_{2i}), $i = 1, 2, ..., n$ are the observations of a random sample from

BPGL($\lambda_t, v, \phi_1, \phi_2$). Then, the log likelihood function $L(\beta)$ is given as:

$$\sum_{i=1}^{n} y_{1i} \log \phi_1 + \sum_{i=1}^{n} y_{2i} \log \phi_2 - (\sum_{i=1}^{n} y_{1i} + \sum_{i=1}^{n} y_{2i}) \log(\lambda_t + \phi_1 + \phi_2)$$

$$+ \sum_{i=1}^{n} (\log(\Gamma(v + x_{1i} + x_{2i} - 1)) + \log((v - 1)(\lambda_t + \phi_1 + \phi_2 + 1) + y_{1i} + y_{2i})$$

$$- \log(y_{1i}!) - \log(y_{2i}!)) \tag{7.24}$$

where $\beta = (\lambda_t, v, \phi_1, \phi_2)$. The MLEs of β is obtained by maximizing (7.24) using numerical methods with the help of statistical software.

7.2.2 Simulation of Parameters of BPGL Distribution

The estimates obtained using the MLE method are analysed through a simulation study in order to check the small and large sample performance of BPGL distribution. So we took N=1000 replications each of sample sizes $n = 50, 100, 200, 400, 800$ for two sets of parameters ($\lambda_t = 0.2, v = 1.1, \phi_1 = 0.15, \phi_2 = 0.2$) and ($\lambda_t = 0.5, v = 1.4, \phi_1 = 0.8, \phi_2 = 0.9$) and hence bias and mean square errors (mse) of parameters were calculated. As the sample size is increasing, bias and mse of each parameter is decreasing.

7.2.3 The Bivariate INAR (1) Process with Paired BPGL Innovations

Let $Y_t = (Y_{t,1}, Y_{t,2}); t = 1, 2, \ldots$ defines a BINAR (1) process as

$$\begin{aligned} Y_{t,1} &= \rho_1 * Y_{t-1,1} + R_{t,1} \\ Y_{t,2} &= \rho_2 * Y_{t-1,2} + R_{t,2}; \quad \rho_i \in (0,1), i = 1, 2. \end{aligned} \tag{7.25}$$

Now suppose the innovation vector $R_t = (R_{t,1}, R_{t,2})$ have the BPGL($\lambda_t, v, \phi_1, \phi_2$) and $R_{t,k}$ is independent from $Y_{s;j}, j = 1, 2$ for each t and $s < t$. Also innovations are independent from the counting series in binomial thinning operator $*$. Then the resulting $Y_t = (Y_{t,1}, Y_{t,2}; t = 1, 2)$ will have BINAR (1) with BPGL innovation denoted by BINAR (1) BPGL and below mentioned proposition defines some of its properties.

Proposition 3. *Suppose the bivariate random vector $Y_t = (Y_{t,1}, Y_{t,2}); t = 1, 2, \ldots$ follows the BINAR (1) BPGL, then*

1. for $i = 1, 2$, the mean, variance and Fisher index of dispersion of $Y_{t,i}$,

$$E(Y_{t,i}) = \frac{(v\lambda_t + v - \lambda_t)\phi_i}{(1 - \rho_i)\lambda_t(\lambda_t + 1)} \tag{7.26}$$

$$Var(Y_{t,i}) = \frac{\phi_i}{1 - \rho_i^2} \left(\rho_i \left(\frac{v}{\lambda_t} - \frac{1}{(\lambda_t + 1)} \right) + \frac{v}{\lambda_t^2(\lambda_t + 1)} \right.$$
$$\left. - \frac{1}{\lambda_t(\lambda_t + 1)^2} + \phi_i \left(\frac{v}{\lambda_t^2} - \frac{1}{(\lambda_t + 1)^2} \right) \right) \tag{7.27}$$

$$DI(Y_{t,i}) = 1 + \frac{\phi_i\left(v(\lambda_t + 1)^2 - \lambda_t^2\right)}{\lambda_t(\lambda_t + 1)(\rho_i + 1)((v - 1)\lambda_t + v)} \qquad (7.28)$$

implying each of $Y_{t,i}$ is over-dispersed marginally.

2. conditional mean and variance of components of the process for $i = 1, 2$,

$$E(Y_{t,i} \mid Y_{t-1,i}) = \rho_i Y_{t-1} + \frac{\phi_i(1 + (v - 1)(\lambda_t + 1))}{\lambda_t(1 + \lambda_t)} \qquad (7.29)$$

and

$$\text{Var}(Y_{t,i} \mid Y_{t-1,i}) = \rho_i(1 - p_i)\rho_{t-1} + \phi_i\left(\frac{v}{\lambda_t^2(\lambda_t + 1)} - \frac{1}{\lambda_t(\lambda_t + 1)^2}\right)$$
$$+ \phi_i^2\left(\frac{v}{\lambda_t^2} - \frac{1}{(\lambda_t + 1)^2}\right) \qquad (7.30)$$

3. covariance of $Y_{t,1}$ and $Y_{t,2}$,

$$\text{Cov}(Y_{t,1}, Y_{t,2}) = \frac{\phi_1\phi_2\left(v(\lambda_t + 1)^2 - \lambda_t^2\right)}{(1 - \rho_1\rho_2)\lambda_t^2(\lambda_t + 1)^2}, \qquad (7.31)$$

4. The conditional joint pmf of the process

$$P(y_t \mid y_{t-1}) = \sum_{k=0}^{u}\sum_{s=0}^{v} z_1(k)z_2(s)P(R_{t,1} = y_{t,1} - k, R_{t,2} = y_{t,2} - s) \quad (7.32)$$

where $u = min(y_{t,1}; y_{t-1,1})$, $v = min(y_{t,2}; y_{t-1,2})$,

$$z_1(k) = \binom{y_{t-1,1}}{k} v_1^k (1 - p_1)^{y_{t-1,1} - k},$$

$$z_2(s) = \binom{y_{t-1,2}}{s} v_2^s (1 - p_2)^{y_{t-1,2} - s}$$

and $P(R_{t,1} = y_{t,1} - k, R_{t,2} = y_{t,2} - s)$ is given by substituting x_1 with $y_{t-1,1} - k$ and x_2 with $y_{t-1,2} - s$ in (7.14).

Also the stationary condition for the model is that $0 < p_i < 1$ (Mamode Khan et al. [2020]) and $E(Y_{t,i})$, $V(Y_{t,i})$ for $i = 1, 2$ and $\text{Cov}(Y_{t,1}, Y_{t,2})$ does not depend on t and $V(Y_{t,i})$ is finite under the conditioned mentioned.

7.2.4 Estimation of Parameters of BINAR (1) BPGL

Suppose $\{Y_t, t = 1, 2, ..., n\}$ is a random sample of size n taken from BINAR (1) BPGL process. Here, the Conditional Maximum Likelihood (CML) approach is used to estimate the parameters $\Theta = (\lambda_t, v, \phi_1, \phi_2, \rho_1, \rho_2)$ of the BINAR (1) BPGL process. The conditional log likelihood function of the BINAR (1) BPGL is

$$\ell(\Theta) \quad = \quad \sum_{t=2}^{n} \log \left[P\left(y_t \mid y_{t-1}\right) \right] \tag{7.33}$$

where $\Theta = (\lambda_t, v, \phi_1, \phi_2, \rho_1, \rho_2)$ is the unknown parametric vector to be estimated and $P\left(y_t \mid y_{t-1}\right)$ is given by (7.32). The cml estimates are obtained by maximizing (7.33). Here optim function and fdHess in R software is used to obtain the cml estimates, observed information matrix and hence standard errors (SE) of estimates of the parameters in BINAR (1) BPGL process.

7.2.5 Simulation Study for BINAR (1) BPGL Process

The CML estimates obtained for the unknown parameters of BINAR (1) BPGL is assessed through a simulation study. Hence, N=1000 samples each of sizes n=25,50,100 are taken for two sets of parametric values $(\lambda_t = 0.2, v = 1.1, \phi_1 = 1, \phi_2 = 2; \rho_1 = 0.5, \rho_2 = 0.2)$ and $(\lambda_t = 0.2, v = 1.1, \phi_1 = 1, \phi_2 = 2, \rho_1 = 0.5, \rho_2 = 0.2)$. For each n, bias and mses were calculated.
As the sample size is increasing, bias and mse of each parameter is decreasing.

7.3 The Sarmanov Bivariate Poisson Generalized Lindley Distribution and Corresponding BINAR (1) Process

Sarmanov [1966] introduced Sarmanov family. Suppose Y_1 and Y_2 are two random variables each with pmf $P(Y_1 = y_1)$ and $P(Y_2 = y_2)$ with supports defined on $A_1 \subseteq \mathbb{R}$ and $A_2 \subseteq \mathbb{R}$, respectively. Now, let $q_i(y_i), i = 1, 2$ be bounded non-constant functions such that

$$\sum_{y_i=-\infty}^{\infty} q_i(y_i) f_i(y_i) = 0. \tag{7.34}$$

Then, the joint pmf for Sarmanov family,

$$P(Y_1 = y_1, Y_2 = y_2) = P_1(y_1 = y_1) P_2(y_2 = y_2) \left[1 + \omega q_1(y_1) q_2(y_2)\right], \tag{7.35}$$

where the factor $\omega q_1(y_1) q_2(y_2)$ is a measure of departure of the two variables Y_1, Y_2 from independence and ω is a real number that satisfies the condition

$$\left[1 + \omega q_1(y_1) q_2(y_2)\right] \geq 0, \text{ for all } y_1, y_2 \tag{7.36}$$

Depending on choices for the functions $q(x)$, we can derive different cases. Here we use $q_i(y_i) = exp(-y_i) - L_i(1)$ $y_i = 0, 1, \dots$ as mentioned in Ting Lee [1996], where

$L_i(1)$ is the value of the Laplace transform of the marginal distribution evaluated at $s = 1$, that is

$$L(s) = E\left(e^{-sY}\right) = \sum_{y=0}^{\infty} e^{-sy} P(y),$$
(7.37)

where $P(.)$ is the marginal distribution. Hence we derive the Sarmanov bivariate Poisson Generalized Lindley (SPGL) having TPPGL as marginals. The Laplace transform for TPPGLD,

$$
\begin{aligned}
L(s) &= E(e^{-sy}) \\
&= \frac{\lambda_t{}^v}{(1+\lambda_t)^{v+1}} e^{-s}(1 - \frac{e^{-s}}{1+\lambda_t})^{-v}(e^s(2+\lambda_t) - 1) \\
&= \frac{\lambda_t{}^v}{\lambda_t+1} \frac{2+\lambda_t - e^{-s}}{(1+\lambda_t - e^{-s})^v}
\end{aligned}
$$
(7.38)

and at $s = 1$,

$$L(1) = \frac{\lambda_t{}^v}{\lambda_t+1} \frac{2+\lambda_t - e^{-1}}{(1+\lambda_t - e^{-1})^v}$$
(7.39)

Then, joint pmf for Sarmanov bivariate distribution takes the form,

$$
\begin{aligned}
P(Y_1 = y_1, Y_2 = y_2) &= P_1(Y_1 = y_1) P_2(Y_2 = y_2) \\
&\quad [1 + \omega\left(\exp(-y_1) - L_1(1)\right)\left(\exp(-y_2) - L_2(1)\right)],
\end{aligned}
$$
(7.40)

where $L_i(.)$ is the Laplace transform for the ith marginal, $i = 1, 2$.

Proposition 4. *The joint pmf of Y_1 and Y_2 each having TPPGLD marginal is,*

$$
\begin{aligned}
P(Y_1 = y_1, Y_2 = y_2) &= \frac{\lambda_{t1}^v \Gamma(y_1+v-1)}{\Gamma(v) y_1! (\lambda_{t1}+1)^{y_1+v+1}} (y_1 + (v-1)(\lambda_{t1}+2)) \times \\
&\quad \frac{\lambda_{t2}^v \Gamma(y_2+v-1)}{\Gamma(v) y_2! (\lambda_{t2}+1)^{y_2+v+1}} (y_2 + (v-1)(\lambda_{t2}+2)) \times \\
&\quad (1 + \omega(e^{-Y_1} - \frac{\lambda_{t1}^v}{(\lambda_{t1}+1)e^{1-v}} \frac{e(2+\lambda_{t1})-1}{(e(1+\lambda_{t1})-1)^v})(e^{-Y_2} - \frac{\lambda_{t2}^v}{(\lambda_{t2}+1)e^{1-v}} \frac{e(2+\lambda_{t2})-1}{(e(1+\lambda_{t2})-1)^v})),
\end{aligned}
$$
(7.41)

where $y_1, y_2 = 0, 1, 2, ..., \lambda_{t1}, \lambda_{t2} > 0, v > 1$ and ω satisfies 7.36.

Therefore, the mean and variance of $Y_i, i = 1, 2$,

$$E(Y_i) = \frac{1 + (v-1)(\lambda_{ti}+1)}{\lambda_{ti}(1+\lambda_{ti})}$$
(7.42)

$$\text{Var}(Y_i) = \frac{v(1+\lambda_{ti})}{\lambda_{ti}^2} - \frac{2+\lambda_{ti}}{(1+\lambda_{ti})^2}.$$
(7.43)

The covariance between Y_1 and Y_2,

$$\text{Cov}(Y_1, Y_2) = \omega u_1 u_2 \tag{7.44}$$

where $u_i = \text{E}(Y_i e^{-Y_i} - Y_i L(1))$. For SPGL,

$$
\begin{aligned}
u_i &= \frac{\lambda_{ti}^{v}(1 - v + e(v(2 + \lambda_{ti}) - \lambda_{ti} - 1))}{e^{1-v}(1 + \lambda_{ti})(e(\lambda_{ti} + 1) - 1)^{v+1}} \\
&\quad - \frac{1 + (v-1)(\lambda_{ti} + 1)}{\lambda_{ti}(1 + \lambda_{ti})} \frac{\lambda_{ti}^{v}}{\lambda_{ti} + 1} \frac{2 + \lambda_{ti} - e^{-1}}{(1 + \lambda_{ti} - e^{-1})^{v}} \\
&= \frac{(1 - e)\left(v(\lambda_{ti} + 1)^2(e(\lambda_{ti} + 2) - 1) - \lambda_{ti}(\lambda_{ti} + 2)(e(\lambda_{ti} + 1) - 1)\right)}{((\lambda_{ti} + 1)^2(e(\lambda_{ti} + 1) - 1)^{v+1})e\lambda_{ti}^{v}} \tag{7.45}
\end{aligned}
$$

7.3.1 Estimation of Parameters of SPGL Distribution

The MLE method is used for the estimation of parameters. Let $(Y_{1i}, Y_{2i}), i = 1, 2, ..., n$ be the observations of a random sample from $\text{SPGL}(\lambda_{t1}, \lambda_{t2}, v, \omega)$. Then the log likelihood function,

$$U(\lambda) = \sum_{i=1}^{n} \log(P(Y_1 = y_{1i}, Y_2 = y_{2i})), \tag{7.46}$$

where $\lambda = (\lambda_{t1}, \lambda_{t2}, v, \omega)$ and $P(Y_1 = y_{1i}, Y_2 = y_{2i})$ is the pmf of SPGL defined in Equation (7.41). Equation (7.24) had to be maximized to find estimates for λ.

7.3.2 Simulation for SPGL Distribution

The mle obtained for the unknown parameters of SPGL is assessed through a simulation study. Hence N=1000 samples each of sizes n=50,100, 200, 400, 800 are taken for two sets of parametric values ($\lambda_{t1} = 0.5, \lambda_{t2} = 0.4, v = 1.5, \omega = 0.9$) and ($\lambda_{t1} = 1.4, \lambda_{t2} = 1.7, v = 2, \omega = 0.1$). For each n, bias and mses were calculated. As the sample size is increasing, bias and mse of each parameter is decreasing.

7.3.3 BINAR (1) Process with Paired SPGL Innovations

Define the bivariate random vector $Y_t = (Y_{t,1}, Y_{t,2}); t = 1, 2$ having BINAR (1) process as in Section 7.2.3 but the innovation vector $R_t = (R_{t,1}, R_{t,2})$ possess $\text{SPGL}(\lambda_{t1}, \lambda_{t2}, v, \omega)$ and the assumptions mentioned in section 7.2.3 holds. The following proposition mentioned explains some of its properties.

Proposition 5. *Suppose the bivariate random vector* $Y_t = (Y_{t,1}, Y_{t,2}); t = 1, 2$ *follows the BINAR (1) SPGL, then*

1. for i=1,2, the mean, variance and Fisher index of dispersion of $Y_{t,i}$,

$$E(Y_{t,i}) = \frac{1 + (v-1)(\lambda_{ti}+1)}{(1-\rho_i)\lambda_{ti}(1+\lambda_{ti})} \tag{7.47}$$

$$Var(Y_{t,i}) = \frac{v(\lambda_{ti}+1)^2(\lambda_{ti}+\lambda_{ti}\rho_i+1) - \lambda_{ti}^2(\lambda_{ti}+\lambda_{ti}\rho_i+\rho_i+2)}{\lambda_{ti}^2(\lambda_{ti}+1)^2(1-\rho_i^2)} \tag{7.48}$$

$$DI(Y_{t,i}) = 1 + \frac{1-v}{((v-1)\lambda_{ti}+v)(1+\rho_i)} + \frac{1}{\lambda_{ti}(1+\rho_i)}$$
$$+ \frac{1}{(\lambda_{ti}+1)(1+\rho_i)} \tag{7.49}$$

implying $Y_{t,i}, i = 1,2$ are over-dispersed.

2. conditional mean and variance of components of the process for i=1,2,

$$E(Y_{t,i} \mid Y_{t-1,i}) = \rho_i Y_{t-1} + \frac{(1+(v-1)(\lambda_{ti}+1))}{(1-\rho_i)\lambda_{ti}(1+\lambda_{ti})} \tag{7.50}$$

and

$$Var(Y_{t,i} \mid Y_{t-1,i}) = \rho_i(1-\rho_i)X_{t-1} + \frac{v(1+\lambda_{ti})}{\lambda_{ti}^2} - \frac{2+\lambda_{ti}}{(1+\lambda_{ti})^2} \tag{7.51}$$

3. covariance of $R_{t,1}$ and $R_{t,2}$,

$$Cov(R_{t,1}, R_{t,2})$$
$$= \omega \frac{(1-e)\left(v(\lambda_{t1}+1)^2(e(\lambda_{t1}+2)-1) - \lambda_{t1}(\lambda_{t1}+2)(e(\lambda_{t1}+1)-1)\right)}{((\lambda_{t1}+1)^2(e(\lambda_{t1}+1)-1)^{v+1})e\lambda_{t1}^v}$$
$$\times \frac{(1-e)\left(v(\lambda_{t2}+1)^2(e(\lambda_{t2}+2)-1) - \lambda_{t2}(\lambda_{t2}+2)(e(\lambda_{t2}+1)-1)\right)}{((\lambda_{t2}+1)^2(e(\lambda_{t2}+1)-1)^{v+1})e\lambda_{t2}^v} \tag{7.52}$$

4. The covariance of two components of the BINAR (1) SPGL of Y_t is obtained as

$$Cov(Y_{t,1}, Y_{t,2}) = \frac{\omega}{(1-\rho_1\rho_2)}$$
$$\times \frac{(1-e)\left(v(\lambda_{t1}+1)^2(e(\lambda_{t1}+2)-1) - \lambda_{t1}(\lambda_{t1}+2)(e(\lambda_{t1}+1)-1)\right)}{((\lambda_{t1}+1)^2(e(\lambda_{t1}+1)-1)^{v+1})e\lambda_{t1}^v}$$
$$\times \frac{(1-e)\left(v(\lambda_{t2}+1)^2(e(\lambda_{t2}+2)-1) - \lambda_{t2}(\lambda_{t2}+2)(e(\lambda_{t2}+1)-1)\right)}{((\lambda_{t2}+1)^2(e(\lambda_{t2}+1)-1)^{v+1})e\lambda_{t2}^v} \tag{7.53}$$

   5. *The conditional joint pmf of BINAR (1) SPGL can be obtained by using Equation (7.32) except that $P(R_{t,1} = r_1, R_{t,2} = r_2)$ is swapped by Equation (7.41).*

Also the stationary condition for the model is that $0 < p_i < 1$ and $E(Y_{t,i})$, $V(Y_{t,i})$ for $i = 1, 2$ and $Cov(Y_{t,1}, Y_{t,2})$ do not depend on t and $V(Y_{t,i})$ is finite under the conditioned mentioned.

7.3.4   Estimation of Parameters of BINAR (1) SPGL

We use the method of CML to estimate the parameters. Suppose $\{Y_t, t = 1, 2, ..., n\}$ is a random sample of size n taken from BINAR (1) SPGL process. Then the conditional log likelihood function is such that,

$$\ell'(\Theta') \quad = \quad \sum_{t=2}^{n} \log\left[P\left(y_t \mid y_{t-1}\right)\right] \tag{7.54}$$

where $\Theta' = (\lambda_{t1}, \lambda_{t2}, v, \omega, \rho_1, \rho_2)$ is the unknown parameteric vector to be estimated and $P(y_t \mid y_{t-1})$ is obtained by substituting the joint pmf of $(R_{t,1}, R_{t,2})$ by (7.41) in (7.32). The CML estimates are obtained by maximizing (7.54). Here optim function and fdHess in R software is used to obtain the cml estimates, observed information matrix and hence standard errors (SE) of estimates of the parameters in BINAR (1) SPGL process.

7.3.5   Simulation Study for BINAR (1) SPGL Process

The cml estimates obtained for the unknown parameters of BINAR (1) SPGL is assessed through a simulation study. Hence N=1000 samples each of sizes n=25,50,100 are taken for two sets of parametric values $(\lambda_{t1} = 0.7, \lambda_{t2} = 0.6, v = 1.1, \omega = 0.2, \rho_1 = 0.1, \rho_2 = 0.7)$ and $(\lambda_{t1} = 0.1, \lambda_{t2} = 0.2, v = 1.5, \omega = 0.8, \rho_1 = 0.4, \rho_2 = 0.9)$. For each n, bias and mses were calculated. As the sample size is increasing, bias and mse of each parameter is decreasing.

7.4   Empirical Study

In this section, the empirical importance of the proposed two BINAR (1) processes are proved using a real dataset. The dataset we have used is on the criminal records of drug activities (CDRUGS) and shooting activities (CSHOTS) in the 12th police car beat in Pittsburgh for period from January 1990 to December 2001 downloaded from the Pittsburgh police departments in the file PghCarBeat.csv (available from: http://www.forecastingprinciples.com/index.php/data.). The mean(variances) of CDRUGS and CSHOTS is 5.1736 (13.1794) and 5.7569 (14.2412), respectively, prove clear over-dispersion.

Now we prove the applicability of the proposed BINAR (1) models by comparing it with some other BINAR (1) models by means of goodness of fit criterion. Hence we compare BINAR (1) BPGL, BINAR (1) SPGL with bivariate Poisson weighted exponential (BINAR (1) BPWE) and bivariate Sarmanov Poisson weighted exponential (BINAR (1) SPWE) by Sajjadnia et al. [2021], (BINAR (1) BP) and BINAR (1) Negative Binomial (BINAR (1) NB) by Pedeli and Karlis [2011], and BINAR (1) Poisson Lindley (BINAR (1) PL) by Mamode Khan et al. [2020]. In Table 7.1 estimates of the parameters, Log-Likelihood(L), AICs, BICs and RMSEs of both the series for all the models described above is calculated. The RMSEs represent the sum of squared differences between true values and one step conditional expectations. As Ristic and Popovic [2019] suggested, the standardized Pearson residuals is calculated for checking the accuracy of the model BINAR (1) SPGL. It is observed that they are uncorrelated and they have means -0.0084, -0.0106 and variances 1.0593, 0.9632.

7.5 Concluding Remarks

Two novel bivariate distributions namely, the Basic BPGL and SPGL are introduced in this chapter. Their mathematical properties are derived. The respective model parameters are estimated using the MLE approach. The simulation results based on these novel bivariate models yield consistent estimates under both distributions. Moreover, BINAR (1) BPGL and BINAR (1) SPGL processes are constructed with these paired innovations. The parameters of the BINAR (1)s are thence estimated using the method of CMLE. The small and large sample performances of these BINAR (1)s are studied and the simulated mean results confirm the consistency of the estimates. These BINAR (1)s are thereon applied to analyze the Pittsburg data and the application results reveal that BINAR (1) SPGL provides better model adequacy measures than some other recently proposed BINAR (1) models. Hence, the BINAR (1)s with Poisson generalized Lindley can be considered among the commendable bivariate time series models and compete against the existing ones in the literature. Their applications are also subject to the nature of the data.

Bivariate Distribution Methods

Model	Estimates	-L	AIC	BIC	RMSEs
BINAR (1) SPGL	$\lambda_{t1} = 0.7870$ $\lambda_{t2} = 0.6257$ $v = 2.8658$ $\rho_1 = 0.4076$ $\rho_2 = 0.3198$ $\omega = 0.9851$	713.6356	1439.2711	1457.0900	3.0358 3.4641
BINAR (1) BPGL	$\lambda_t = 2.9967$ $v = 4.9462$ $\phi_1 = 2.3869$ $\phi_2 = 3.1247$ $\rho_1 = 0.3574$ $\rho_2 = 0.2470$	724.8362	1461.6725	1479.4914	3.0754 3.4975
BINAR (1) BPWE	$\mu_1 = 2.3271$ $\mu_2 = 3.4914$ $\rho_1 = 0.4883$ $\rho_2 = 0.3344$	750.664	1509.3279	1521.2071	3.0087 3.4742
BINAR (1) SPWE	$\tau_1 = 0.3515$ $\tau_2 = 0.2949$ $\rho_1 = 0.4952$ $\rho_2 = 0.3973$ $\omega = 0.8949$	723.2353	1456.4706	1471.3196	3.0029 3.4516
BINAR (1) BP	$\lambda_1 = 2.644$ $\lambda_2 = 3.7131$ $\phi = 0.2852$ $\rho_1 = 0.4243$ $\rho_2 = 0.3249$	765.8323	1541.6647	1556.5138	3.0413 3.4654
BINAR (1) NB	$\lambda_1 = 2.6630$ $\lambda_2 = 3.6450$ $\beta = 0.3163$ $\rho_1 = 0.427$ $\rho_2 = 0.3045$	728.7059	1467.4119	1482.2610	3.0907 3.4857
BINAR (1) PL	$\lambda_{t1} = 0.5464$ $\lambda_{t2} = 0.4494$ $\rho_1 = 0.431$ $\rho_2 = 0.3479$ $\omega = -0.9306$	720.7707	1451.5414	1466.3905	3.022 3.4563

TABLE 7.1

Estimates and goodness of fit statistic of BINAR (1) models

8

Extension of BINAR (1) to BINAR(p) Proces

Another aspect that still intrigues researchers, especially in the analytics realm, is to understand the coherence between the COVID-19 infection and death cases in Mauritius via some advanced bivariate statistical methodologies. In today's ages, by studying both COVID-19 infection and death series, authorities will be able to assess the effectiveness of the vaccines and understand the behavior of the COVID-19 pandemic in specific countries, especially when the number of COVID-19 infected cases is increasing though the rise in the share of fully vaccinated Mauritian people and the emergence of more variants with vivid propagation rate. By understanding the patterns in the COVID-19 infected and death series, one can also easily undertake pro-active measures regarding scenarios like need for adequate ventilators in COVID-19 specialised hospitals, number of burials per day and most importantly, identify the factors or measures not reaping expected positive outcomes in containing SARS-CoV-2.

Also, as we already know, patients with huge comorbidities are prone to COVID-19 infection and related deaths thus considering that Mauritius has a high prevalence of Non-Communicable Disease, such research findings allow health authorities to work towards maintaining a good population health via proper health strategies. In fact, the SARS-CoV-2 can cause acute myocardial injury and chronic damage to the cardiovascular system in the long run. Blood tests have shown an increase in troponin levels (cardiac biomarkers) in patients affected with SARS-CoV-2. These cardiac issues have definitely impacted on the quality of life in these patients. Mortality with cardiac problems has also increased significantly since the emergence of this novel pandemic. Some studies have shown that beyond the first 30 days after infection, individuals with COVID-19 are at an increased risk of incident cardiovascular disease, including cerebrovascular disorders, dysrhythmias, ischemic and non-ischemic heart disease, pericarditis,myocarditis, heart failure and thromboembolic disease. These risks and burdens were evident even among individuals who were not hospitalized during the acute phase of the infection and increased in a graded fashion according to the care setting during the acute phase (non-hospitalized,hospitalized and admitted to intensive care). A growing number of studies suggest many COVID-19 survivors experience some type of heart damage, even if they did not have underlying heart disease and were not sick enough to be hospitalized. Troponin level has been shown to be elevated in these patients affected with COVID-19 pneumonia. Therefore, in this book, many health-related complications have been induced into the model to assess its relationship with regard to the COVID-19 infected and deaths cases in Mauritius. Worldwide many such researches have been conducted but Mauritius, despite having

DOI: 10.1201/9781003677451-8

high prevalence of NCDs, was neglected. This book thus shall surely add value to the existing medical related research. As a yardstick for good policy making, such research findings will allow concerned health authorities to assess the fatality rate related to COVID-19, as well as assess the vulnerability of the population and consequently, take remedial actions like more vaccine with high effective rate roll-outs among patients with complex medial history.

8.1 Brief on Bivariate Integer-Valued Auto-Regressive Model

As already mentioned earlier, the BINAR (1) model starts from the pioneered work of Pedeli and Karlis [2011], where the authors introduced the bivariate integer-valued process of auto-regressive nature by considering two classical thinning-based INAR (1) series described in McKenzie [1986], Al Osh and Alzaid [1987], McKenzie [1988] with some cross correlated innovation series. In particular, the BINAR (1) was introduced under strict stationary conditions, that is, the thinning or auto-correlation parameters in both series lie in the interval $[0, 1)$ and the corresponding paired innovations were marginally distributed with constant distributional parameters or under time-constant covariates Pedeli and Karlis [2013b]. In these BINAR (1)s, the paired innovation terms induce the cross dependence between the two series. Overall, such BINAR (1) produces a diagonal matrix of auto-correlation parameters.

Furthermore, in Pedeli and Karlis [2011], the BINAR (1) has been constructed under the simple binomial thinning with constant coefficient [Steutel and Van Harn, 1979] and paired innovations were assumed to follow the popular bivariate Poisson [Kocherlakota and Kocherlakota, 1992] and bivariate Negative-Binomial models [Marshall and Olkin, 1990]. Other works in the same direction include the papers by Pedeli and Karlis [2013b] where the paired innovations were constructed from copulas. As aforementioned, these BINAR (1)s were developed under strict stationarity conditions. While, Mamode Khan et al. [2016], Mamode Khan et al. [2018] and others [Jowaheer et al., 2018] proposed some extensions of the stationary BINAR (1)s by allowing the paired innovations to follow some bivariate models under time-dependent moments. The latter BINAR (1)s can thus be used to model real-life time series that exhibit some non-stationarity in the presence of time-varying covariates. As for the inferential procedures, the CML method was used to estimate the model parameters efficiently, and proves to yield estimates with lower standard errors and lesser bias than Yule-Walker or Method of Moments.

Some other important extensions include the BINAR (1) studied by Ristic et al. [2012], Nastic et al. [2016], Popovic et al. [2018a,b] with geometric marginal distribution based on negative binomial thinning operator. Scotto et al. [2014a] proposed the bivariate binomial autoregressive model suitable for modeling count data with a finite range of counts. Yu et al. [2020] generalized the bivariate constant coefficient model discussed by Karlis and Pedeli [2013] to the bivariate random coefficient model. The non-diagonal BINAR (1) has also been considered in Karlis and Pedeli

[2013]. However, the estimation procedures in these extended BINAR (1)s are rather cumbersome. Thus in this book, we propose to extend the diagonal BINAR (1) to the BINAR(p) process based on the binomial thinning procedure. Thereon, we assess the proposed BINAR(p) to simulated data and the COVID-19 data and suggest improvements. These are described in the forthcoming sections.

8.2   The BINAR(p) Process and Some Properties

By letting $Y_{t,i}$ be the random count observation at the tth time point for the ith variable with $y_{t,i}$ as a realization of $Y_{t,i}$, where $i \in \{1,2\}$, the BINAR(p) can be expressed as:

$$Y_{t,1} = \rho_{11} * Y_{t-1,1} + \rho_{21} * Y_{t-2,1} + \rho_{31} * Y_{t-3,1} + \cdots + \rho_{p1} * Y_{t-p,1} + R_{t,1} \quad (8.1)$$

$$Y_{t,2} = \rho_{12} * Y_{t-1,2} + \rho_{22} * Y_{t-2,2} + \rho_{32} * Y_{t-3,2} + \cdots + \rho_{p2} * Y_{t-p,2} + R_{t,2} \quad (8.2)$$

where, $*$ is the binomial thinning operator. The paired innovation term represented by $(R_{t,1}, R_{t,2})$, follows some bivariate distribution denoted by BVD such that $(R_{t,1}, R_{t,2})$ $\sim$ BVD $(r_{t,1}, r_{t,2}; \mu_{t,1}, \mu_{t,2}, v_1, v_2, \phi)$ with $\mu_{t,i}$ and v_i are the marginal mean and dispersion parameter of $R_{t,i}$ respectively and $Cov(R_{t,1}, R_{t,2}) = \phi$. The other assumptions include $Cov(Y_{t-j,i}, R_{t,i}) = 0$, $Cov(Y_{t-j,1}, R_{t,2}) = 0$, $Cov(Y_{t-j,2}, R_{t,1}) = 0$ for $j \in Z^+$.

From these assumptions and following the Lemma 2.1 in Du and Li [1991],

$$E(Y_{t,i}) = \sum_{j=1}^{p} \rho_{ji} E(Y_{t-j,i}) + E(R_{t,i})$$

$$V(Y_{t,i}) = \sum_{j=1}^{p} \rho_{ji}(1 - \rho_{ji}) E(Y_{t-j,i}) + \sum_{j=1}^{p} \rho^2_{ji} V(Y_{t-j,i})$$

$$+ 2 \sum_{j=1}^{p} \sum_{j<j'} \rho_{ji} \rho_{j'i} Cov(Y_{t-j,i}, Y_{t-j',i}) + V(R_t, i)$$

Alternatively,

$$V(Y_{t,i}) = E(Y_{t,i}) + \sum_{j=1}^{p} \rho^2_{ji}[V(Y_{t-j,i}) - E(Y_{t,i})]$$

$$+ 2 \sum_{j=1}^{p} \sum_{j<j'} \rho_{ji} \rho_{j'i} Cov(Y_{t-j,i}, Y_{t-j',i}) + [V(R_{t,i}) - E(R_{t,i})]$$

Note that, $Cov(Y_{t,1}, Y_{t,2}) = \sum_{j=1}^{p} \sum_{k<j}^{p} Cov(\rho_{j1} * Y_{t-j,1}, \rho_{k2} * Y_{t-k,2}) + Cov(R_{t,1}, R_{t,2})$ Thus, the Fisher index of dispersion, that is, the ratio $V(Y_{t,i})$ to $E(Y_{t,i})$ is greater than 1, provided the $R_{t,i}$ can be marginally Poisson or extra-Poisson distributed. On the other hand, the $Cov(Y_{t,1}, Y_{t,2})$ depends on the survival parts of the two series and the

paired innovations. This indicates that we can relax the assumption of cross correlation on the innovation terms and yet the pair of observations $(Y_{t,1}, Y_{t,2})$ are correlated. Furthermore, by assuming $\tilde{Y}_t = (Y_{t,1}, Y_{t,2})$, the joint conditional density function
is given by:

$$f_{\tilde{Y}_t}(\tilde{y}_t \mid \mathscr{F}_{t,1}, \mathscr{F}_{t,2}) = \sum_{k=0}^{g_1}\sum_{s=0}^{g_2} f_1(k) f_2(s) \text{BVD}(y_{t-k,1}, y_{t-s,2}; \mu_{t,1}, \mu_{t,2}, v_1, v_2, \phi)$$

with $\mathscr{F}_{t,1} = (y_{t-1,1}, y_{t-2,1}, \ldots, y_{t-p,1})$ and, $\mathscr{F}_{t,2} = (y_{t-1,2}, y_{t-2,2}, \ldots, y_{t-p,2})$, $g_1 = min(y_{t,1}, y_{t-1,1}, \ldots, y_{t-p,1})$, $g_2 = min(y_{t,2}, y_{t-1,2}, \ldots, y_{t-p,2})$. The auxiliary functions
are

$$f_1(k) = \sum_{j_1+j_2+\ldots+j_p \leq k} \binom{y_{t-1,1}}{j_1}\binom{y_{t-2,1}}{j_2}\ldots\binom{y_{t-p,1}}{j_p}$$
$$\times \rho_{11}(1-\rho_{11})^{y_{[t-1,1]}-j_1} \times \rho_{21}(1-\rho_{21})^{y_{[t-2,1]}-j_2}$$
$$\times \ldots \times \rho_{p1}(1-\rho_{p1})^{y_{[t-p,1]}-j_p}$$

$$f_2(s) = \sum_{l_1+l_2+\ldots+l_p \leq s} \binom{y_{t-1,2}}{l_1}\binom{y_{t-2,2}}{l_2}\ldots\binom{y_{t-p,2}}{l_p}$$
$$\times \rho_{12}(1-\rho_{12})^{y_{[t-1,2]}-l_1} \times \rho_{22}(1-\rho_{22})^{y_{[t-2,2]}-l_2}$$
$$\times \ldots \times \rho_{p2}(1-\rho_{p2})^{y_{[t-p,2]}-l_p}$$

The conditional likelihood equation to estimate the vector of model parameters indicated by θ is then given by

$$L(\tilde{\theta} \mid \tilde{Y}) = \prod_{t=1}^{T} f(\tilde{Y}_t \mid \mathscr{F}_{t,1}, \mathscr{F}_{t,2})$$

From Billingsley [1961b], Bu et al. [2008], the $(\hat{\theta} - \theta) \sim \text{Normal}(0, I^{-1}(\hat{\theta}))$.
The above BINAR (1) model can accommodate for the covariate specifications,
that is, assuming a common vector of p covariates for the ith variate given by
$X_t = [x_{t,1}, x_{t,2}, \ldots, x_{t,p}]^T$, and with regression coefficients for the ith variate given by
$\beta_i = [\beta_{i,1}, \beta_{i,2}, \ldots, \beta_{p,i}]$, the $\mu_{t,i}$ can be obtained by $\mu_{t,i} = \exp(X_t^T \beta_i)$.

8.2.1   Sub-Models

Several sub-models can be developed from Equations (8.1) and (8.2), by assuming
fewer terms in either $Y_{t,1}$ or $Y_{t,2}$. Likewise, we can assume a BINAR(a,b), with $a < p$
and $b < p$ and with different innovation distributions that may be cross correlated
(via copulas) Pedeli and Karlis [2013b,c] or not correlated. More interestingly, the
$Y_{t,2}$ may be allowed to depend on some $Y_{t-k,1}$ for $k < p$ as described in Pedeli and
Karlis [2013ab], and this can lead to a non-diagonal cross correlation structure. The

construction of such sub-models depend on the nature of the problem we are investigating and on computational issues that may arise with the full BINAR(p) process.

Another solution to the above can be obtained by modeling only the priority random variable, say $Y_{t,2}$, but conditioned on the past occurrences of the other variable, that is, $Y_{t-k,1}$. Likewise, we can assume $Y_{t,1}$ to represent the number of COVID-19 cases and $Y_{t,2}$ as the priority random variable for the number of COVID-19 deaths at the tth time point. It is likely that the number of deaths $Y_{t,2}$ depends on the past number of cases, and in such situation, we can assume an Integer-valued Generalized Auto-regressive Conditional Heteroscedastic (IN-GARCH) equation of order p, q and k of the form

$$Y_{t,2} \mid Y_{t-1,2}, Y_{t-2,2}, \ldots Y_{t-p,2}, Y_{t-1,1}, \ldots, Y_{t-k,1} \sim D(\lambda_t, \omega) \tag{8.3}$$

where D is the probability model

$$
\begin{aligned}
&E(Y_{t,2} \mid, Y_{t-1,2}, Y_{t-2,2}, \ldots Y_{t-p,2}, Y_{t-1,1}, \ldots, Y_{t-k,1}) \\
&= \alpha_0 + \sum_{i=1}^{p} \alpha_i Y_{t-i,2} + \sum_{j=1}^{q} \beta_j \mu_{t-j,1} + \sum_{s=1}^{k} \xi_s * Y_{t-s,1}
\end{aligned}
\tag{8.4}
$$

or for large $Y_{t-i,1}$ like countries having huge number of cases, we can assume

$$
\begin{aligned}
&E(Y_{t,2} \mid, Y_{t-1,2}, Y_{t-2,2}, \ldots Y_{t-p,2}, Y_{t-1,1}, \ldots, Y_{t-k,1}) \\
&= \alpha_0 + \sum_{i=1}^{p} \alpha_i Y_{t-i,2} + \sum_{j=1}^{q} \beta_j \mu_{t-j,1} + \sum_{s=1}^{k} \xi_s * Y_{t-s,1}
\end{aligned}
\tag{8.5}
$$

where $\alpha_i, \beta_j, \xi_k > 0$, where the $*$ operator is substituted by the multiplication sign. Such model representation is analogous to Zhu [2009], Zhu and Li [2009], Zhu [2012a,b,c], Weiß et al. [2021]. However, in Equation 8.3, the effects on $Y_{t,1}$ are not computed.

8.3 Numerical Illustrations

This section assesses the performance of the proposed BINAR(p) model, assuming that the pair of innovations $\{R_{t,1}, R_{t,2}\}$ is bivariate Poisson with parameters $\{\mu_1, \mu_2, \phi\}$, i.e, $v_1 = v_2 = 1$. From Karlis and Ntzoufras [2003], the bivariate Poisson is expressed as

$$
\begin{aligned}
&P_{(R_{t,1}, R_{t,2})}(r_{t,1}, r_{t,2}) \\
&= \exp\{-(\mu_1 + \mu_2 + \phi)\} \frac{\mu_{1_{t,1}}^r}{r_{t,1}!} \frac{\mu_{2_{t,2}}^r}{r_{t,2}!} \sum_{k=0}^{\min(r_{t,1}, r_{t,2})} \binom{r_{t,1}}{k} \binom{r_{t,2}}{k} k! \left(\frac{\phi}{\mu_1 \mu_2}\right)^k
\end{aligned}
$$

Simulation	Model	Parameter	True Value	Simulated Value	STD Error
50	BINAR 2	$\rho_1 1$	0.2	0.188	0.136
		$\rho_2 1$	0.1	0.078	0.091
		$\rho_1 2$	0.3	0.305	0.156
		$\rho_2 2$	0.4	0.359	0.127
		μ_1	1	1.101	0.504
		μ_2	1	1.103	0.692
		ϕ	0.5	0.493	0.378
100	BINAR 2	$\rho_1 1$	0.2	0.184	0.095
		$\rho_2 1$	0.1	0.112	0.096
		$\rho_1 2$	0.3	0.311	0.102
		$\rho_2 2$	0.4	0.38	0.085
		μ_1	1	1.019	0.335
		μ_2	1	1.052	0.456
		ϕ	0.5	0.472	0.244
200	BINAR 2	$\rho_1 1$	0.2	0.191	0.066
		$\rho_2 1$	0.1	0.098	0.063
		$\rho_1 2$	0.3	0.302	0.072
		$\rho_2 2$	0.4	0.385	0.067
		μ_1	1	1.03	0.246
		μ_2	1	1.055	0.325
		ϕ	0.5	0.486	0.195
500	BINAR 2	$\rho_1 1$	0.2	0.197	0.043
		$\rho_2 1$	0.1	0.097	0.044
		$\rho_1 2$	0.3	0.302	0.041
		$\rho_2 2$	0.4	0.298	0.039
		μ_1	1	1.0015	0.152
		μ_2	1	1.03	0.197
		ϕ	0.5	0.494	0.112

TABLE 8.1
Simulated mean results for BINAR (2) Process

and can be simulated with the desired parameters using the *bivpois* package down-loadable. We consider the cases where $p = 2$ and 3 and for sample sizes $T = 50$, 100, 200, and 500 and the combinations: $\rho_{11} = 0.2$, $\rho_{21} = 0.1$, $\rho_{31} = 0.05$ and $\rho_{12} = 0.3$, $\rho_{22} = 0.4$, $\rho_{32} = 0.03$, $\mu_1 = \mu_2 = 1$, and $\phi = 0.5$. The simulation experiments are executed for 500 replications at each combination, and the following tables record the simulated mean estimates of the different model parameters. The

Simulation	Model	Parameter	True Value	Simulated Value	STD Error
50	BINAR 3	$\rho_1 1$	0.2	0.183	0.074
		$\rho_2 1$	0.1	0.092	0.061
		$\rho_3 1$	0.05	0.07	0.054
		$\rho_1 2$	0.3	0.221	0.072
		$\rho_2 2$	0.4	0.384	0.068
		$\rho_3 2$	0.03	0.057	0.050
		μ_1	1	0.897	0.302
		μ_2	1	1.271	0.357
		ϕ	0.5	0.606	0.211
100	BINAR 3	$\rho_1 1$	0.2	0.193	0.101
		$\rho_2 1$	0.1	0.1	0.084
		$\rho_3 1$	0.05	0.061	0.067
		$\rho_1 2$	0.3	0.26	0.114
		$\rho_2 2$	0.4	0.377	0.098
		$\rho_3 2$	0.03	0.058	0.071
		μ_1	1	1.002	0.391
		μ_2	1	1.18	0.594
		ϕ	0.5	0.479	0.315
200	BINAR 3	$\rho_1 1$	0.2	0.194	0.074
		$\rho_2 1$	0.1	0.087	0.061
		$\rho_3 1$	0.05	0.056	0.054
		$\rho_1 2$	0.3	0.288	0.072
		$\rho_2 2$	0.4	0.386	0.068
		$\rho_3 2$	0.03	0.047	0.050
		μ_1	1	1.005	0.302
		μ_2	1	1.012	0.357
		ϕ	0.5	0.523	0.211
500	BINAR 3	$\rho_1 1$	0.2	0.199	0.044
		$\rho_2 1$	0.1	0.097	0.045
		$\rho_3 1$	0.05	0.053	0.036
		$\rho_1 2$	0.3	0.294	0.043
		$\rho_2 2$	0.4	0.395	0.041
		$\rho_3 2$	0.03	0.036	0.030
		μ_1	1	0.99	0.186
		μ_2	1	1.014	0.220
		ϕ	0.5	0.511	0.131

TABLE 8.2
Simulated mean results for BINAR (3) process

results from Tables 8.1 and 8.2 demonstrate that the simulated mean estimates of the model parameters are consistent with the respective true values. As the sample size increases, the standard errors under each combinations, decreases. However, the execution times for the BINAR (2) for the large $T = 300, 500$ are significantly huge, and as we move to the BINAR (3), the codes become very heavily time consuming even for $T = 50$. We consider alternative minimization solutions like the *nlm* and *DEoptim* but the BINAR for higher orders $(p > 1)$ become very time consuming. In fact, for the BINAR(3) cases, the computations become terribly slow and take many days to yield the outputs. In some practical applications, it is worth to consider separately the two INAR processes and induce the inter-relation by some common explanatory variables in the mean function $\mu_{t,i}$. That is, we first obtain the solution set for $Y_{t,1}$: $(\rho_{11}, \rho_{21}, \ldots, \rho_{p1}, \mu_1, \nu_1)$ and then obtain the set for $Y_{t,2}$: $(\rho_{12}, \rho_{22}, \ldots, \rho_{p2} \mu_2, \nu_2)$. For optimizing the INAR(p), the reader may refer to the papers by Pedeli et al. [2015a], Joe [2019] and the recent paper by Soobhug et al. [2022]. We use this approach in the simulated series above and fit two INAR (2) models for the case $p = 2$ and two INAR (3) for the case $p = 3$, respectively. The important remark is the model estimates are still consistent with some slight bias due to the slightly higher standard errors which are as expected. Thus, for the forthcoming section on the application, we may opt for fitting separate INARs in case the BINAR may be practically unfeasible.

8.4   Application to the COVID-19 Series in Mauritius

Refer to Figure 1.3, for the COVID-19 data on the daily number of new infection cases and deaths from 21 March 2020 to 25 April 2021, which aggregate to 400 pairs of observations. Other information on these data includes the COVID-19 Stringency index which indicates the severity of the measures imposed by the Mauritian government to control the propagation of the SARS-CoV-2 virus. The remaining variables include population characteristics such as the median age of the population, population density and comorbidities related indicators such as the cardiovascular death rate, the diabetes prevalence and proportion of smokers among others. By using the *adf.test* in the library *tseries*, the Dickey-Fuller statistic gives -3.8374 with a p-value of 0.0177 for the COVID-19 new infection cases series signifying, time stationarity. On the other hand, for the COVID-19 new death series, the Dickey-Fuller is -6.044 with p-value of less than 0.01, which also proves that the death series is stationary. An insight on the basic statistics for the COVID-19 new infection and death series are tabulated as in Table 1.1 in Chapter 1. From Figure 1.3, the autocorrelation values decrease as the number of lags increases and hence the time series with auto-regressive nature is justified. The PACF plots show a number of significant lags. Further to the number of non-convergent simulations for fairly $p > 2$ reported, we suggest a simple modified bivariate INAR process to analyze the series. In fact, the cross correlation test at lag 1 executed from the *corrtest* package in R gives

Parameters	ρ_{11}	ρ_{21}	ρ_{22}	ω_1	ω_2	c	λ
Estimates	0.1115	0.2109	0.0080	0.0417	0.9974	0.1141	2.7053
sd	0.0456	0.0571	0.0030	0.0204	0.4045	0.0531	0.8314
p-value	0.0146	0.0002	0.0077	0.0409	0.0137	0.0316	0.0011

TABLE 8.3
Estimates of the new BINAR process: No covariates

Listing 8.1: Test of Cross correlation

```
Tests  for  zero  cross-correlation  of  cases  and  deaths  (Mauritius)

| Lag|     CC|  Lag|         t | p-value
|---:|-----:|---------------|--------
|  -1|  0.258|-1  |     1.887|    0.059|
|   0|  0.134| 0  |     2.689|    0.007|
|   1|  0.119| 1  |     2.387|    0.017|
```

Thus, the BINAR at lag 1 is a suitable model for the data. Since the COVID-19 death depends on being infected, we propose a new BINAR model of the form:

$$Y_{t,1} = \rho_{11} * Y_{t-1,1} + R_{t,1} Y_{t,2} = \rho_{21} * Y_{t-1,2} + \rho_{22} * Y_{t-1,1} + R_{t,2} \qquad (8.6)$$

where $Y_{t,1}$ measures the number of COVID-19 new infection cases at the tth time point and $Y_{t,2}$ is the corresponding number of COVID-19 new death cases. To avoid computational failures with the use of complex PWE and PGLD and their ZI versions, as reported in Chapter 5, we start this first work on bivariate processes, by assuming that $R_{t,1} \sim \text{NegBinom}(c, \omega_1)$, where *NegBinom* is the NB distribution with size c and probability ω_1 and $R_{t,2}$ is the ZIP with mean λ and probability of zeros denoted by ω_2.

8.5 Results and Comments

8.5.1 Models with No Covariates

We first fit the Mauritius COVID-19 data using the new BINAR process in Equation (8.6) assuming there are no covariate effects, and the results are given as: The AIC from the above analysis is 1400.352 and we can conclude that the estimates from the new BINAR model are significant and fits the data well with a satisfactory AIC. We can fit the using a corresponding zero-inflated Poisson IN-GARCH (λ_t, ω_2) as described in 8.5 for the death series assuming $p = 1, q = 0, k = 1$, that is,

$E(Y_{t,2}|Y_{t-1,1}, Y_{t-1,2}) = \lambda_t = \alpha_0 + \alpha_1 Y_{t-1,2} + \xi_1 Y_{t-1,1}$ and $\omega_2 = \frac{\exp(\beta)}{1+\exp(\beta)}$. The conditional likelihood is given

$$L(\theta|Y_2) = \prod_{t=2}^{T} f(Y_{t,2} \mid Y_{t-1,1}, Y_{t-1,2}) \tag{8.7}$$

where $f(.) = dzip(y_{t,2}, \lambda_t, \omega_2)$. The *dzip* is the zero-inflated Poisson density function in the package *ZIM* in R. The results from the IN-GARCH process are shown in the table below as: These results again demonstrate and confirm from Listing 1 that there is a significant dependence between the COVID-19 new deaths and new infection cases, with reference to the estimate of ξ_1. Hence, the need to fit these COVID-19 data via the new BINAR model and with at least one common covariate is justified. The next subsection explores this feature.

8.5.2   The BINAR Model with Covariates

From Equation 8.6, we assume $(Y_{t,1})$ that the most influential covariate is the COVID-19 Stringency Index variable for the COVID-19 in Mauritius [Mamode Khan et al., 2020b, Soobhug et al., 2022]. From the negative binomial definition, we may allow $\omega_1 = \frac{c}{c+\mu_1}$ with $\mu_1 = \exp(\beta_{0,1} \times \text{stringency})$, where the stringency variable is coded as 0 for stringency index below 15 and 1 for greater than 15. The same stringency index is included in the COVID-19 death series link predictor equation to induce the cross correlation between the cases and deaths as follows. For the COVID-19 death series, we let

$$\lambda = \text{stringency}^{\beta_{0,2}} \times \text{median age}^{\beta_1} \times \text{cardiovascular death rate}^{\beta_2}$$
$$\times \text{diabetes rate}^{\beta_3} \times \text{smokers proportion}^{\beta_4}$$

The stringency variable is included in both cases and death series. The estimates of the above variables are tabulated in Table 8.5 below: With log likelihood value 1406.286 and AIC 2834.572, the stringency estimates: stringency (1) and stringency (2) for the COVID-19 new infection cases and deaths are highly significant. For Mauritius, during the first phase, due to school closures, travel restrictions, institution of quarantine centers and other strict sanitary restrictions, the stringency index was near

Parameter	α_0	α_1	ξ_1	β
Estimates	0.0093	0.0278	0.0046	1.0011
sd		0.0001	0.0141	0.0012	0.0001
p-value		0.0000	0.0487	0.0001	0.0000

TABLE 8.4
Estimates from the IN-GARCH (1,0,1) process

Parameters	stringency (1)	stringency (2)	med. age	cardio death	diabetes	smokers prop.
Estimates	3.7573	0.5351	0.0121	0.1128	0.0713	0.1215
sd	0.0363	0.1246	0.0032	0.0301	0.0243	0.0321
p-value	0.0000	0.0000	0.0002	≤ 0.0002	0.0033	0.0002

TABLE 8.5
Regression estimates from the new BINAR process

15 and this resulted in a decline in the number of COVID-19 new infection cases and consequently, fewer COVID-19 related deaths cases. In 2021, the stringency index was towards the lower side due to lessening of sanitary restrictions especially opening of borders. This provoked more COVID-19 infected cases of asymptomatic nature and more deaths cases, despite the speedy vaccine roll-outs and mandatory wearing of face masks in public places. Conclusively, the stringency index has been computed based on nine metrics and by observing the COVID-19 situation in Mauritius, we can say that if all nine metrics are implemented in Mauritius with full vigor, then we will be able to capture its full effect that is the number of COVID-19 infection cases and related death cases will be more rapidly and successfully curbed. Conversely, if only a few metrics are imposed, the effect will be very partial. The age factor and comorbidities like the cardiovascular diseases, diabetes and smoking status are significant to the COVID-19 deaths, with cardio-death as the most significant. There are cardiac affections related to COVID-19, but until now no cause of COVID-19 death has been reported due to cardiac issues. We did however found cardiovascular damages like significant increase in cardiac bio-marker, troponin, for those patients with mild, moderate and severe COVID-19 pneumonia. Diabetes and smoking are two risk factors for cardiovascular disease and hence confirm their significance.

It is noticeable that the number of COVID-19 new infection cases and deaths were almost consistent with the change in the stringency index, except some very few cases of both infected and deaths occurred with a large stringency index. These confirm the stationary nature of the data.

8.6 Concluding Remarks

The proposed BINAR process offers huge flexibility in terms of choice of the innovation distribution to suit data features such as excess of zeros, covariate specification (both time-varying and time-independent covariates) and can handle quite long sequences of data. In the specific application to the COVID-19 series in Mauritius, several risk factors such as cardio, smoking, and stringency index was detected significant and hence this implies that forthcoming health and policy measures have to be geared towards improving the COVID-19 stringency index and increase regular monitoring on elder patients, especially those suffering from cardiovascular problems.

However, this area of research do have some limitations for instance, the COVID-19 pandemic is a novel one thus anonymized micro information for many countries, are not yet available, possibly due to lack of timely recording. This make it difficult to access some important variables like medical history of COVID-19 patients, weather factors, and other which may affect the propagation of the SARS-CoV-2 virus and emerging variants [Refer to Mamode Khan et al. [2020a], Soobhug et al. [2022] for some interesting real-life applications studies.

On the other hand, we attempted a BINAR (4) with the covariates and noticed that the algorithm used an extensively huge amount of time. The main research question is now whether the high ordered BINAR models can be used to analyze COVID-19 series in other countries facing huge number of COVID-19 cases and deaths. In fact, by considering non-dependent innovation terms, the application of the new BINAR to such countries become less computationally cumbersome. However, in case one of the series consists of huge values, an INAR model coupled with an auto-regressive moving average process model through using copula can be an alternative solution.

9

Summary and Future Directions

9.1 Summary

The different features of the COVID-19 series in Mauritius builds the foundations of this book. Several models under the family of high-ordered integer-valued autoregressive processes, most precisely, INAR(p) processes with different Poisson-mixtures innovations and their corresponding zero-inflated parts under different thinning mechanisms, were considered thoroughly while taking into account the effect of time-varying covariate specifications and non-stationarity. As discussed earlier in Chapter 2, such processes are highly suitable to model over-dispersed time series with higher orders. In Chapter 3, the INAR(p) processes with different Poisson mixtures innovations were mounted and tested via simulation experiments. The simulation experiments allowed us to conclude that ZI-NB, ZI-PWE, and ZI-PGLD, under the GB thinning, were able to produce less biased estimates and satisfactory standard errors. On the other, we also observe smooth execution of the binomial thinning under any scenarios, without any simulation failures and this allowed us to further explore the best suited innovation distribution under Chapters 5 to 8. Moreover, both CLS and CML approaches are shown to provide fairly consistent estimates, but with CML yielding slightly lesser biased estimates under both Binomial and Generalized Binomial thinnings. Some computational failures were noticed with the CML approach in particular with the CMP, PT, PIG, PL and WCG innovations. The results under Chapter 3, motivated us to move to Chapter 4 whereby different ZI-models under the GB and NB thinnings were applied to the COVID-19 series in Mauritius. Here, the non-stationary INAR (7) with ZI-NB, ZI-PWE and ZI-PGLD innovations under the GB thinnings were shown to provide lower AICs though its complex algorithm were really time consuming to execute. Note that working under the NB thinning proved not only to be computationally challenging but the AIC that it reaps were not comparatively very satisfactory. Conversely, the estimates from the Generalized binomial thinning show less biasedness, but this may be subject to the quality of the data as well thus GB should be used restrictively. Given the satisfactory performance of the ZI-PWE and ZI-PGLD distributions after the ZI-NB, under Chapter 5, an attempt to further explore them, while catering for periodic aspect observed in the COVID-19 new infection and death series of South Africa and Mauritius, were conducted. Different periodic INAR processes with Poisson-mixtures models and their ZI versions were implemented and assessed under the non-stationary setting.

Here, again the periodic INAR processes with ZI-NB outperformed the other models. As mentioned earlier, PWE and PGLD were recently emerged and considering their performance under univariate framework, we explored the bivariate processes of the PWE and PGLD under Chapters 6 and 7, respectively. We explored extensively two models namely the bivariate of order one with PWE innovation and the bivariate of order one with PGLD innovation under the stock markets data series and the monthly crime data of Pittsburgh, respectively. These two series worked well for bivariate INAR (1) processes but this also motivated us to extend it to BINAR(p) processes so that high-ordered integer-valued series can be studied too. Chapter 8 brings forward the properties of BINAR(p) processes and assess its applicability on the COVID-19 new infection and death series of Mauritius under both the stationary and non-stationary setting. Here, various time varying covariates were used namely the COVID-19 stringency index which has proved to be highly significant in reducing the number of new COVID-19 infection cases in Mauritius and in South Africa, the age factor and comorbidities like the cardiovascular diseases, diabetes and smoking status of COVID-19 patients. All these proved to be significant however, the takeaways from this study is the utmost need to widen the list of metrics considered for the computation of the COVID-19 stringency index. For instance, latent variable like behavior of the population with regard to the newly established COVID-19 specific legislations, adherence to vaccination roll-outs, can be induced into the computation of this index. This work also came up with reliable forecasts results which will surely help authorities to come up with more preparedness and remedial strategies.

9.2   Future Directions

This work explored thoroughly the COVID-19 new infection and death series focused on Mauritius, however, one aspect that requires much attention is the regional or spatial disparity observed in the propagation of the COVID-19 virus and the detection of variants. In fact, initially the World Health Organisation (WHO) describes four transmission scenarios for COVID 19: No cases: Countries/ territories/ areas with no cases; Sporadic cases: Countries/ territories/ areas with one or more cases, imported or locally detected; Clusters of cases: Countries/ territories/ areas experiencing cases, clustered in time, geographic location, and/or by common exposure; and finally, Community transmission: Countries/ territories/areas experiencing larger outbreaks of local transmission, defined through an assessment of factors, which make us understand that a spatial association of the COVID-19 infections exist and assessing the propagation of the COVID-19 virus while catering for the geographical feature is important. Europe and the United States of America are the most affected countries with more than 1 million daily confirmed cases. Most importantly, let's focus on the spatial aspect in Europe. All European countries have been affected by COVID-19 which means that there is a high spatial auto-correlation among the neighboring European countries. Now, within each European country, say France, there can be additional variables that could affect the propagation of the coronavirus like the strictness of the sanitary legislations, the environmental factors and others. Similar case is for the United States of America and its dependencies as well as for African countries. Most African countries have a negligible share of confirmed COVID-19 infection cases, which again confirm the presence of a spatial aspect among neighboring countries or even among states.

It can be observed that the neighboring cities of Wuhan namely Qichun county in Huanggang prefecture, Tianjin, Chongqing, Jiangxi have been affected by COVID-19 and Kang et al. [2020] via the Moran's I spatial statistics confirmed that there is a significant spatial dependence of COVID-19 infections amongst the these adjacent cities of Wuhan. Yet another interesting distribution whereby the geographical feature should be studied is the emergence of variants.

We clearly observed similar spreading pattern of the Omicron variants among neighboring American, Asian, and African countries.

Under this particular real-life data application and many others like road accident analysis [Pirdavani et al., 2013, Shafabakhsh et al., 2017], it is clear that spatial aspect is important but how to cater for this?

Under statistical terms, these COVID-19 series by regions or cities, are spatial count observations and are often influenced by similar events occurring in neighboring areas or by site-variant or invariant factors like environmental factors and others. Another interesting feature of spatial count is the presence of over-dispersion, and due to the dependence on similar occurrences in neighboring countries, the spatial observations may exhibit some serial autocorrelation with its previous lagged observation resulting into a spatial temporal model. Also, as described earlier, spatial count observations are influenced by site-variant and invariant factors thus stationarity or non-stationarity can occur. Finally but most importantly, under the analysis of the spatial data, several parameters need to be estimated which requires a robust estimation procedure. We thus proposes to address these features under another class emanating from the family of integer-valued autoregressive models - the spatial integer-valued autoregressive models (SINAR).

9.2.1 Brief on Spatial Integer-Valued Auto-Regressive (SINAR) Model for the 2-Dimensional Spatial Data

Pickard [1980], Basu and Reinsel [1993], and Martin [1996] proposed the following spatial auto-regressive model for the continuous data, based on the first-order neighboring unilateral structure:

$$Y_{i,j} = \rho_1 Y_{i-1,j} + \rho_2 Y_{i,j-1} + \rho_3 Y_{i-1,j-1} + R_{ij} \qquad (9.1)$$

where R_{ij} is the white noise with mean=0 and variance=σ^2. However, as elaborated in earlier section, most spatial integer-valued time series are of discrete nature, requiring more exploration. Here, we make reference to the discrete analog of Equation (9.1), that is termed as the SINAR (1) process, proposed by Ghodsi et al. [2012]:

$$Y_{i,j} = \rho_1 * Y_{i-1,j} + \rho_2 * Y_{i,j-1} + \rho_3 * Y_{i-1,j-1} + R_{ij} \qquad (9.2)$$

where $*$ is the binomial thinning operator and R_{ij} is assumed to be i.i.d with mean μ_t and σ^2. Here, an important aspect is that Ghodsi et al. [2012] assumed strict stationary conditions when formulating the Equation (9.2), that is, $Cov(Y_{ij}, Y_{i-k,j-l}) = Cov(Y_{ij}, Y_{i+k,j+l})$, $K \geq 1$ and $l \geq 1$. The marginal means and covariances can be obtained by solving a quadratic characteristic equation—$Cov(Y_{ij}, Y_{i-k,j-l}) = C\lambda^k \eta^{-l}$, following Basu and Reinsel [1993] and Mickens [1990].

9.2.2 Limitations and Way Forward

The limitations regarding the SINAR (1) and also the rationale for future works are elaborately provided below:

1. In the current literature, mostly reference is made to the first-order Spatial Integer-Valued Autoregressive (SINAR (1,1)) model in the stationarity setting using binomial thinning with Poisson innovation term. However, in real-life applications, with the huge over-dispersion phenomenon, non-stationarity, excess zeros, and harmonic effects observed in most series, the SINAR (1,1) with Poisson term model may fall short in catering for these features. Therefore, we propose, just like in Chapters 4 and 5 of this book, to extend the SINAR (1,1) process with popular discrete Generalized Poisson models or mixtures, such as the COM-Poisson, Poisson-Gamma (marginally,the Negative Binomial), Poisson-Lindley, Poisson Inverse-Gaussian, Poisson Tweedie, Weighted Cosine-Geometric, Modified-Geometric, PWE and the most recent PGLD models and their associated zero-inflated versions in the error specification, R_{ij}, under both the stationary and non-stationary settings. Here, the CML estimation approach is preferred because as shown in Chapter 3 and 4, the CML outperformed CLS with lower standard errors. As a rule of parsimony and to avoid computational failures, the focus will be on using binomial thinning operator.

2. The SINAR (1) process assumed strict stationary but as discussed earlier, during spatial analysis, many areas or cells can be influenced by site-variant or invariant factors. For instance among the European countries, for the case of COVID-19, Akter et al. [2022], Jablonska et al. [2021], Kissler et al. [2021], and Weaver et al. [2012] concluded that predictors like country-specific demographical, environmental characteristics, health system and preventive measures, influenced the spread of the main four variants of concern namely Alpha, Beta, Gamma and Delta in that region. In this context, it is important to induce time-varying variants in the first-order integer-valued SINAR process. It is thus proposed to first come up with a flexible non-stationary SINAR (1) models with the regular Rook, Bishop, and Queen lattices which shall cater for intriguing features in the COVID-19 series like over-dispersion, serial autocorrelation and others, as well.

3. It is also important to account for the time dependence in spatial analysis. The spatial observation Y_{ij} is serially autocorrelated with its previous observations which, under the spatial context, gives rise to two spatial correlation structures : the serial autocorrelation and the spatial autocorrelation structure. Under the existing SINAR models, spatio-temporal correlation is not catered thus we propose to come up with a spatio-temporal autoregressive (STAR) model for the integer-valued case which is indexed by space and time simultaneously. It is also proposed to work on extending the model by having moving average (MA) terms and hence STARMA (or even STARIMA or seasonal STARIMA models).

4. Considering the emergence of high number of integer-valued time series with high orders, it becomes imperative to extend the SINAR (1) process to a SINAR(p) process with different Poisson-mixtures innovation distributions. However, this work requires much time and is complex.

These research works will be an addition to the current literature thus the efforts to explore the family of integer-valued autoregressive models are ongoing.

Appendices

So far, the simulation results and the boxplots illustrate consistent estimates of the different parameters, but are still incomplete, especially for the $T = 500$ and for $p = 7$ as the algorithms become very time consuming and we also intend to run some simulations with the bivariate Negative-Binomial innovations described in the *binegbin* package in R [Kirkpatrick and Neale, 2015]. Besides, in the application, we intend to study the relationship between the number of COVID-19 deaths series and the number of new COVID-19 infected patient with acute symptoms admitted to the hospital. This is so because due to the ongoing vaccination campaigns, the number of asymptomatic cases are climbing thus for better assessment of the severity of the situation, the number of new COVID-19 infected patient with acute symptoms admitted to the hospital, now seem to be a better indicator. Such information related to some developed countries are available at $http://shiny.webpopix.org/covidix/app3/$.

A.1 Scenario: Poisson innovation with binomial thinning, assuming stationary setting

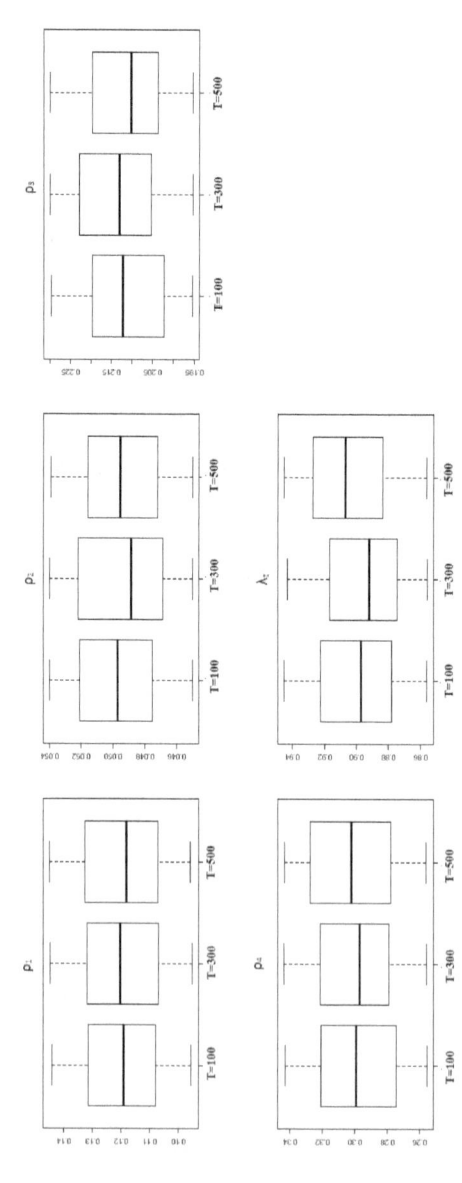

FIGURE A.1

Boxplot for Poisson innovation with binomial thinning, assuming stationary setting

A.2 Scenario: ZI-P innovation with binomial thinning, assuming non-stationary setting

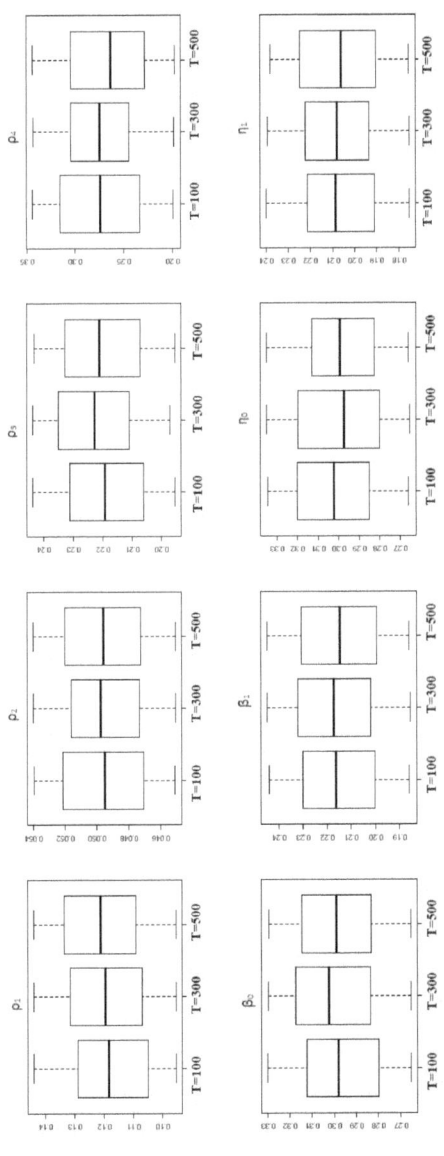

FIGURE A.2
Boxplot for ZI-P innovation with binomial thinning, assuming non-stationary setting

A.3   Scenario: Poisson innovation with Generalized Binomial thinning, assuming non-stationary setting

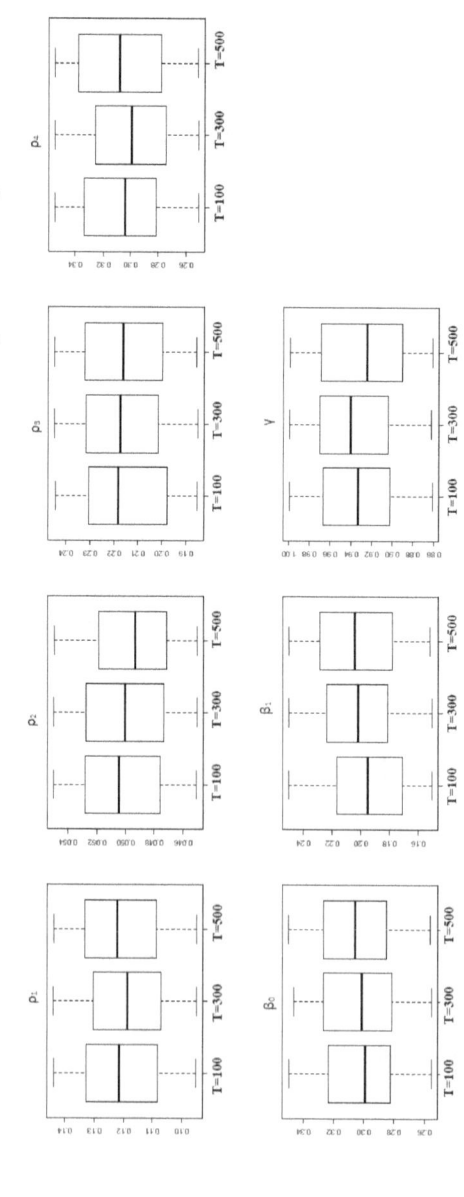

FIGURE A.3
Boxplot for Poisson innovation with Generalized Binomial thinning, assuming non-stationary setting

A.4 Scenario: ZI-P innovation with Generalized Binomial thinning, assuming stationary setting

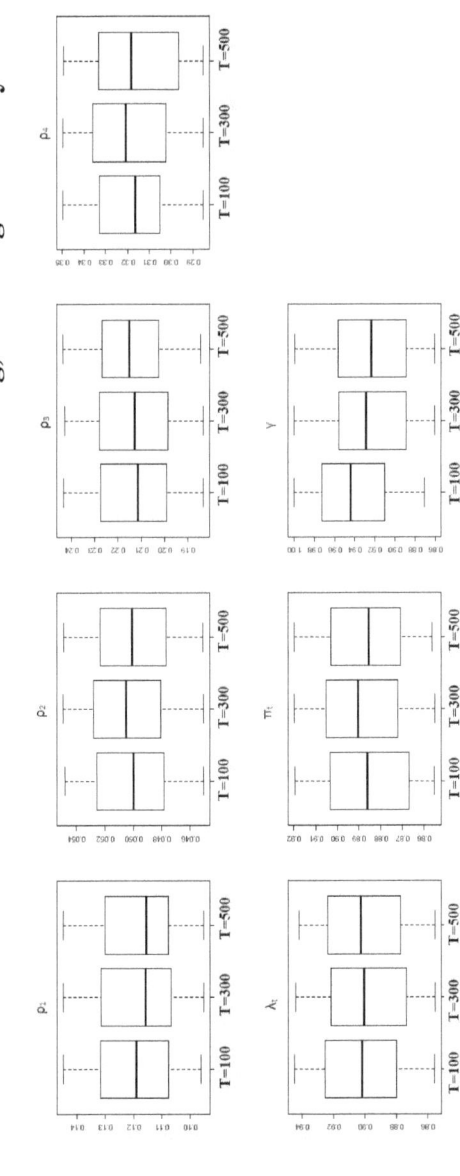

FIGURE A.4
Boxplot for ZI-P innovation with Generalized Binomial thinning, assuming stationary setting

A.5   Boxplots from Some Simulations: BINAR(*p*)

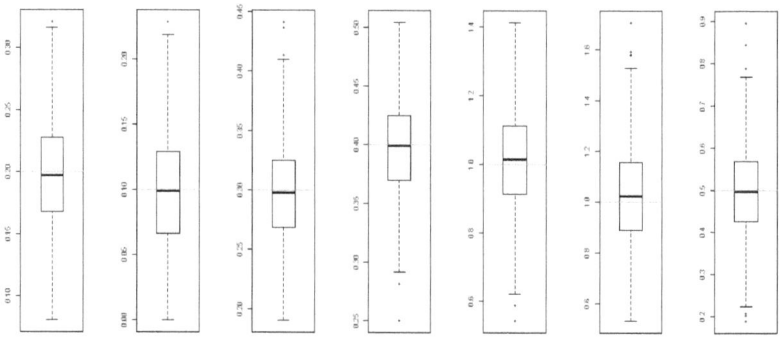

FIGURE A.5
Boxplot of BINAR (2) model for all parameters and T=500

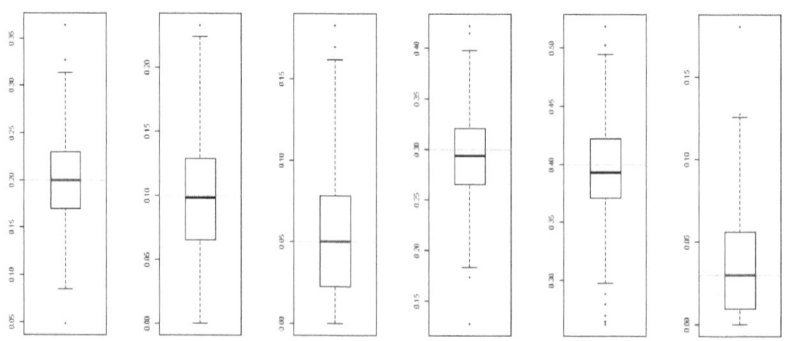

FIGURE A.6
Boxplot of BINAR (3) model for parameters ρ_{11}, ρ_{21}, ρ_{31}, ρ_{12}, ρ_{22}, ρ_{32} and T=500

Bibliography

A. Abouammoh, A. M. Alshangiti, and I. Ragab. A new generalized lindley distribution. *Journal of Statistical Computation and Simulation*, 85(18):3662–3678, 2015.

S. Akter, M.A. Zakia, M. Mofijur, S.F. Ahmed, D.N. Vo, G. Khandaker, and T.M.I. Mahlia. Sars-cov-2 variants and environmental effects of lockdowns, masks and vaccination: A review. *Environ Chem Lett.*, 20(1):141–152, 2022.

M.A. Al Osh. The impact of missing data in a generalized integer-valued autoregression model for count data. *Journal of Biopharmaceutical Statistics*, 19:1039–1054, 2009.

M.A. Al Osh and E-E.A.A. Aly. First order autoregressive time series with negative binomial and geometric marginals. *Communications in Statistics-Theory and Methods*, 21(9):2483–2492, 1992.

M.A. Al Osh and A.A. Alzaid. First-order integer-valued autoregressive process. *Journal of Time Series Analysis*, 8:261–275, 1987.

E. Altun. Weighted-exponential regression model: An alternative to the gamma regression model. *International Journal of Modeling, Simulation, and Scientific Computing*, 10(6):2535–2550, 2019.

E. Altun. A new generalization of geometric distribution with properties and applications. *Communications in Statistics-Simulation and Computation*, 49(3):793–807, 2020.

E-E.A.A. Aly and N. Bouzar. Explicit stationary distributions for some galton-watson processes with immigration. *Communications in Statistics—Stochastic Models*, 10:499–517, 1994.

E-E.A.A. Aly and N. Bouzar. Expectation thinning operators based on linear fractional probability generating functions. *Journal of the Indian Society for Probability and Statistics*, 20:89–107, 2019.

A.A. Alzaid and M.A. Al Osh. An integer-valued pth order autoregressive structure (INAR(p)) process. *Journal of Applied Probability*, 27:314–324, 1990.

A.A. Alzaid and M.A. Al Osh. Some autoregressive moving average processes with generalized poisson marginal distributions. *Annals of the Institute of Statistical Mathematics*, 45:223–232, 1993.

A. Andersson and K. Dimitris. A parametric time series model with covariates for integers in z. *Statistical Modelling*, 14:135–156, 2014.

N. Aries and N. Mamode Khan. On periodic integer-valued moving average (inma(q)) models.

Armstrong and Green. Forecasting principles. 2017. URL http://www.forecastingprinciples.com/index.php/data.

M. Awale, N. Balakrishna, and T.V. Ramanathan. Testing the constancy of the thinning parameter in a random coefficient integer autoregressive model. *Statistical Papers*, 60:1515–1539, 2019.

M. Awale, A.K. Kashikar, and T.V. Ramanathan. Forecasting overdispersed inar(1) count time series with negative binomial marginal. *Communications in Statistics - Simulation and Computation*, pages 1–21, 2021.

H.S. Bakouch and M.M. Ristic. Zero truncated poisson integer valued ar(1) model. *Metrika*, 72:265–280, 2010.

H.S. Bakouch, M. Mohammadpour, and M. Shirozhan. A zero-inflated geometric inar(1) process with random coefficient. *Applications of Mathematics*, 63:79–105, 2017.

H.S. Bakouch, Y. Sunechar, N. Mamode Khan, and V. Jowaheer. A non-stationary bivariate inar(1) process with a simple cross-dependence: Estimation with some properties. *Australian and New Zealand Journal of Statistics*, 62:25–48, 2021.

N. Balakrishna and Chin Diew Lai. Construction of bivariate distributions. In *Continuous Bivariate Distributions*, pages 179–228. Springer, 2009.

W. Bareto Souza. Zero-modified geometric inar(1) process for modelling count time series with deflation or inflation of zeros. *J Time Ser Anal*, 33(6):839–852, 2015.

S. Basu and G.C. Reinsel. Properties of the spatial unilateral first-order arma model. *Advances in Applied Probability*, page 631–648, 1993.

L. Bauwens and D. Veredas. The stochastic conditional duration model: A latent factor model for the analysis of financial durations. *Journal of Econometrics*, 119 (2):381–412, 2004.

BBC. One covid vaccine cuts infection rate in all age groups. *Article*, 2021. URL https://www.bbc.com/news/health-56844220.

M. Bentarzi and N. Aries. On some periodic inarma(p,q) models. *Journal of Communications in Statistics - Simulation and Computation*, 2020. doi: 10.1080/03610918.2020.1780443.

M. Bentarzi and W. Bentarzi. Periodic integer-valued garch(1,1) model. *Communications in Statistics-Simulation and Computation*, 46(2):1167–1188, 2017.

L. Bermudez and D. Karlis. Multivariate inar(1) regression models based on the sarmanov distribution. *Mathematics*, 9:1–505, 2021.

M.A. Billah, M.M. Miah, M.A. Khan, and N. Khan. Reproductive number of coronavirus: A systematic review and meta-analysis based on global level evidence. *PLOS ONE*, 15:1–17, 2020.

P. Billingsley. Statistical inference for markov processes. *University of Chicago Press, Chicago*, 1961a.

P. Billingsley. Statistical methods in markov chains. *University of Chicago Press*, 1961b.

R. Blundell, R. Griffith, and J. Van Reenen. Market share, market value and innovation in a panel of british manufacturing firms. *The Review of Economic Studies*, 66:529–554, 1999.

W.H. Bonat, B. Jorgensen, C.C. Kokonendji, J. Hinde, and C.G.B. Demétrio. Extended poisson-tweedie: properties and regression models for count data. *Statistical Modelling*, 18:24–49, 2017.

M. Boudreault, H. Cossette, D. Landriault, and E. Marceau. On risk model with dependence between interclaim arrivals and claim sizes. *Scandinavian Actuarial Journal*, 5:265–285, 2006.

M. Bourgignon, J. Rodrigues, and M. Santos Neto. Extended poisson inar(1) processes with equidispersion, underdispersion and overdispersion. *Journal of Applied Statistics*, 46:101–118, 2018.

M. Bourguignon. Modelling time series of counts with deflation or inflation of zeros. *Statistics and Its Interface*, 11:631–639, 2018.

M. Bourguignon and K.L.P. Vasconcellos. Improved estimation for Poisson INAR(1) models. *Journal of Statistical Computation and Simulation*, 85:2425–2441, 2015.

G.E.P. Box and G.M. Jenkins. *Time series analysis: Forecasting and control.* 1970.

K. Brannas and P. Johansson. Time series count data regression. *Communications in Statistics : Theory and Methods*, 23:2907–2925, 1994.

K. Brannas and A.M.M.S. Quoreshi. Integer-valued moving average modelling of the number of transactions in stocks. *Applied Financial Economics*, 20:1429–1440, 2010.

T. Brijs, D. Karlis, and G. Wets. Studying the effect of weather conditions on daily crash counts using a discrete time series model. *Accident Analysis and Prevention*, 40(3):1180–1190, 2008.

P.J. Brockwell and R.A. Davis. *Time Series: Theory and Methods*, volume 82(5). 1991.

K. Brännäs and J. Hellström. Generalized integer-valued autoregression. *Econometric Reviews*, 20:425–443, 2001.

K. Brännäs, J. Hellström, and J. Nordström. A new approach to modelling and forecasting monthly guest nights in hotels. *International Journal of Forecasting*, 18: 19–30, 2002.

R. Bu, B.P.M. McCabe, and K. Hadri. Maximum likelihood estimation of higher-order integer-valued autoregressive processes. *Journal of Time Series Analysis*, 29:973–994, 2008.

Q. Bukhari, J. M. Massaro, R. B. D'Agostino, and S. Khan. Effects of weather on coronavirus pandemic. *International journal of environmental research and public health*, 17, 2020. doi: 10.3390/ijerph17155399. URL https://www.ncbi.nlm.nih.gov/pmc/articles/PMC7432279/.

U. Böckenholt. Mixed inar (1) poisson regression models: analyzing heterogeneity and serial dependencies in longitudinal count data. *Journal of Econometrics*, 89: 317–338, 1998.

M.J. Campbell. Time series regression for counts: an investigation into the relationship between sudden infant death syndrome and environmental temperature. *Journal of the Royal Statistical Society. Series A (Statistics in Society)*, 157:191–208, 1994.

M. Cardinal, R. Roy, and J. Lambert. On the application of integer-valued time series models for the analysis of disease incidence. *Statistics in Medicine*, 18:2015–2039, 1999.

Centers for Disease Control and Prevention. Cdc real-world study confirms protective benefits of mrna covid-19 vaccines. 2021. URL https://www.cdc.gov/media/releases/2021/p0329-COVID-19-Vaccines.html.

S. Chan, J. Chu, Y. Zhang, and S. Nadarajah. Count regression models for covid-19. *Physica A: Statistical Mechanics and Its Applications*, 563:125460, 2021.

H. Chang and D.G. Saunders. Predictors of attrition in two types of group programs for men who batter. *J. Fam. Violence*, 17:273–292, 2002.

S. Chatla and G. Shmueli. Efficient estimation of com–poisson regression and a generalized additive model. *Computational Statistics and Data analysis*, 121:71–88, 2018.

C. Chesneau, H.S. Bakouch, T. Hussain, and B. A. Para. The cosine geometric distribution with count data modeling. *Journal of Applied Statistics*, 48:124–137, 2020.

H. Cho, C. Liu, J. Park, and D. Wu. bzinb: Bivariate zero-inflated negative binomial model estimator r package version 1.0.3. 2019. URL https://cran.r-project.org/package=bzinb.

A. Chudik, M. H. Pesaran, and A. Rebucci. Covid-19 time-varying reproduction numbers worldwide: An empirical analysis of mandatory and voluntary social distancing. *medRxiv*, 2021.

A. Chutoo, D. Karlis, N. Mamode Khan, and V. Jowaheer. The unilateral spatial autoregressive process for the regular lattice two-dimensional spatial discrete data. *SORT*, 2021.

K. S. Conceicao, F. Louzada, M. G. Andrade, and E. S. Helou. Zero-modified power series distribution and its hurdle distribution version. *Journal of Statistical Computation And Simulation*, 87:1842–1862, 2017.

H. Cossette, P. Gaillardetz, E. Marceau, and J. Rioux. On two dependent individual risk models. *Insurance: Mathematics and Economics*, 30:153–166, 2002.

D.R Cox. Statistical analysis of time series: Some recent developments. *Scandinavian Journal of Statistics*, 8:93–115, 1981.

J.A. Cranford, R.A. Zucker, J.M. Jester, L.I. Puttler, and H.E. Fitzgerald. Parental alcohol involvement and adolescent alcohol expectancies predict alcohol involvement in male adolescents. *Psychology of addictive behaviors : Journal of the Society of Psychologists in Addictive Behaviors*, 24:386–396, 2010.

Y. Cui and R. Lund. A new look at time series of counts. *Biometrika*, 96:781–792, 2009.

R.B. Davies. Numerical inversion of a characteristic function. *Biometrika*, 60:415–17, 1973.

R.A. Davis and R. Wu. A negative binomial model for time series of counts. *Biometrika*, 96, 2009.

R.A. Davis, W.T.M. Dunsmuir, and Y. Wang. On autocorrelation in a poisson regression model. *Biometrika*, 90:777–790, 2000.

C. D. Desjardins. Evaluating the performance of two competing models of school suspension under simulation-the zero-inflated negative binomial and the negative binomial hurdle. *(Doctoral Dissertation, University of Minnesota)*, 2013.

C. D. Desjardins. Modeling zero-inflated and overdispersed count data: An empirical study of school suspensions. *The Journal of Experimental Education*, 84(3):449–472, 2016.

J.P. Dion, G. Gauthier, and A. Latour. Branching processes with immigration and integer-valued time series. *Serdica, Mathematical Journal*, 21:123–136, 1995.

S. Dobricic, E. Pisoni, L. Pozzoli, R. Van Dingenen, T. Lettieri, J. Wilson, and E. Vignati. Do environmental factors such as weather conditions and air pollution influence covid-19 outbreaks? *Science for Policy Report by the Joint Research Centre (JRC), the European Commission*, 2020. URL `https://ec.europa.eu/jrc`.

P. Doukhan, N. Mamode Khan, and M.H. Neumann. Mixing properties of non-stationary ingarch(1,1) processes. *ALEA, Latin American Journal of Probability and Mathematical Statistics*, 18:401–420, 2021.

J. Du and Y. Li. The integer-valued autoregressive (INAR(p)) model. *Journal of Time Series Analysis*, 12:129–142, 1991.

O. Dyer. Covid-19: Us reports low rate of new infections in people already vaccinated. *BMJ*, 373, 2021. doi: 10.1136/bmj.n1000.

A. H. El Shaarawia, R. Zhu, and H. Joe. Modelling species abundance using the poisson–tweedie family. *Environmetrics*, 22:152–164, 2010.

M. S Eliwa, E. Altun, M. El Dawoody, and M. El Morshedy. A new three-parameter discrete distribution with associated INAR(1) process and applications. *IEEE access*, 8:91150–91162, 2020.

A. Emrah and N. Mamode Khan. Modelling with the novel INAR(1)-PTE process. *Methodology and Computing in Applied Probability*, pages 1–17, 2021.

D.J.G. Farlie. The performance of some correlation coefficients for a general bivariate distribution. *Biometrika*, 47:307–323, 1960.

R. Ferland, A. Latour, and D. Oraichi. Integer-valued garch process. *Journal of Time Series Analysis*, 27:923–942, 2006.

P.R.P. Filho, V.A. Reisen, P. Bondon, M. Ispany, M. M. Melo, and F.S. Serpa. A periodic and seasonal statistical model for non-negative integer-valued time series with an application to dispensed medications in respiratory diseases. *Applied Mathematical Modelling*, 96:545–558, 2021.

K. Fokianos. Count time series models. *Time Series-Methods and Applications*, 30: 315–347, 2012.

J. Franke and T. Rao. Multivariate first-order integer-valued autoregressions. *Technical Report. Math. Dep., UMIST*, 1995.

R.K. Freeland. Statistical analysis of discrete time series with applications to the analysis of workers compensation claims data. 1998.

R.K. Freeland and B.P.M McCabe. Analysis of low count time series data by Poisson autoregression. *Journal of Time Series Analysis*, 25:701–722, 2004.

R. E. Gaunt, S. Iyengar, A.B.O. Daalhuis, and B. Simsek. An asymptotic expansion for the normalizing constant of the conway-maxwell-poisson distribution. *Annals of the Institute of Statistical Mathematics*, 71:163–180, 2016.

G. Gauthier and A. Latour. Convergence forte des estimateurs des parametres d'un processus genar(p). *Annales des Sciences Mathematiques du Quebec*, 18:46–71, 1994.

M.E. Ghitany and D.K. Al Mutairi. Estimation methods for the discrete poisson–lindley distribution. *Journal of Statistical Computation and Simulation*, 79: 1–9, 2009.

Mohamed E Ghitany, Barbra Atieh, and Saralees Nadarajah. Lindley distribution and its application. *Mathematics and Computers in Simulation*, 78(4):493–506, 2008.

A. Ghodsi, M. Shitan, and H. Bakouch. A first-order spatial integer-valued autoregressive sinar(1,1) model. *Communications in Statistics- Theory and Methods*, 41(15):2773–2787, 2012.

E.G. Gladysev. Periodically correlated random sequences. *Soviet Mathematics*, 2: 385–388, 1961.

E. Gomez-Deniz and E. Calderin Ojeda. The discrete Lindley distribution: properties and applications. *Journal of statistical somputation and simulation*, 81(11):1405–1416, 2011.

E. Gomez-Deniz, J. M. Sarabia, and N Balakrishnan. A multivariate discrete poisson-lindley distribution: Extensions and actuarial applications. *ASTIN Bulletin: The Journal of the IAA*, 42(2):655–678, 2012.

E. Gomez Deniz, J.N. Sarabia, and N. Balakrishnan. Multivariate discrete poisson-lindley distribution: extensions and actuarial applications. *ASTIN Bulletin*, 42(2): 655–678, 2012.

E. Goncalves, N. Mendes-Lopes, and F. Silva. Zero-inflated compound poisson distributions in integer-valued garch models. *Statistics*, 50:558–578, 2016.

C. Gouriéroux and J. Jasiak. Heterogeneous inar (1) model with application to car insurance. *Insurance: Mathematics and Economics*, 34:177–192, 2004.

X. Guo, H. Zhang, and Y.P. Zeng. Transmissibility of covid-19 and its association with temperature and humidity. *Research Square*, 1, 2020. doi: 10.21203/rs.3.rs-17715/v1.

E. J. Haas, F. J. Angulo, J. M. McLaughlin, E. Anis, S. R. Singer, F. Khan, N. Brooks, M. Smaja, G. Mircus, K. Pan, J. Southern, D. L. Swerdlow, L. Jodar, Y. Levy, and S. Alroy-Preis. Impact and effectiveness of mrna bnt162b2 vaccine against sars-cov-2 infections and covid-19 cases, hospitalisations, and deaths following a nationwide vaccination campaign in israel: an observational study using national surveillance data. *The Lancet Journal*, 397:1819–29, 2021.

C.M. Hafner and H. Manner. Dynamic stochastic copula models: estimation, inference and applications. *Journal of Applied Econometrics*, 27(2):269–295, 2012.

D.C. Heilbron. Zero-altered and other regression models for count data with added zeros. *Biometrical Journal*, 36(5):531–547, 1994.

J. Hellström. Unit root testing in integer-valued ar(1) models. *Economics Letters*, 70: 9–14, 2001.

J. Huang and F. Zhu. A new first-order integer-valued autoregressive model with bell innovations. *Entropy*, 23:1–17, 2021.

G. Iacobucci. Covid-19: Infections fell by 65% after first dose of astrazeneca or pfizer vaccine, data show. *BMJ*, 373, 2021. doi: 10.1136/bmj.n1068.

M. R. Irshad, C. Chesneau, V. D'cruz, and R. Maya. Discrete pseudo Lindley distribution: Properties, estimation and application on INAR(1) process. *Mathematical and Computational Applications*, 26(4):76, 2021a.

M. R. Irshad, V. D'cruz, R. Maya, and N. Mamode Khan. Inferential properties with a novel two parameter Poisson generalized Lindley distribution with regression and application to INAR(1) process. *(Communicated)*, 2021b.

K. Jablonska, S. Aballea, and M. Toumi. Factors influencing the covid-19 daily deaths' peak across european countries. *Public Health*, 194:135–142, 2021.

P. Jacobs and P. Lewis. Discrete time series generated by mixtures : correlational and runs properties. *Journal of the Royal Statistical Society Series B*, 40:94–105, 1978a.

P. Jacobs and P. Lewis. Discrete time series generated by mixtures ii: asymptotic properties. *Journal of the Royal Statistical Society Series B*, 40:222–228, 1978b.

S.B. Javali and P.V. Pandit. Using zero inflated models to analyze dental caries with many zeroes. *Indian J. Dent. Res.*, 21:480–485, 2010.

M.A. Jazi, G. Jones, and C.D. Lai. First-order integer valued processes with zero inflated poisson innovations. *Journal of Time Series Analysis*, 33:954–963, 2012.

T. Jeewa. Is the mauritian covid-19 vaccine rollout based on a human rights approach? *Opiniojuris.*, page 144, 2021.

M. Jochmann. zic: Bayesian inference for zero-inflated count models, r package version 0.9.1. 2017. URL https://cran.r-project.org/package=zic.

H. Joe. Likelihood inference for generalized integer autoregressive time series models. *Econometrics*, 7:43, 2019.

B. Jorgensen and C.C. Kokonendji. Discrete dispersion models and their tweedie asymptotics. *AStA Advances in Statistical Analysis*, 100:43–78, 2014.

V. Jowaheer and B.C Sutradhar. Analysing longitudinal count data with overdipsersion. *Biometrika*, 89:389–399, 2002.

V. Jowaheer, N.A. Mamode Khan, and Y. Sunecher. A non-stationary binar(1) process with negative binomial innovations for modeling the number of goals in the first and second half: the case study of arsenal football club. *Communication in Statistics-Case Studies, Data Analysis and Applications*, 2:21–33, 2017.

V. Jowaheer, N. Mamode Khan, and Y. Sunechar. A binar(1) time series model with cross-correlated com-poisson innovations. *Communications in Statistics-Theory and Methods*, 47:1133–1154, 2018.

V. Jowaheer, M. Heenaye Mamode Khan, and Y. Sunechar. Road traffic accident data analysis in mauritius using statistical techniques. *Mauritius Research and Innovation Council*, 2019.

J. Juno and A. Wheatley. Mounting evidence suggests covid vaccines do reduce transmission. how does this work? *The Conversation.*, 2021.

D. Kang, D. Choi, H. Kim, and J. Choi. Spatial epidemic dynamics of the covid-19 outbreak in china. *International Journal of Infectious Diseases*, 94:96–102, 2020.

D. Karlis and I. Ntzoufras. Analysis of sports data by using bivariate poisson models. *Journal of the Royal Statistical Society*, 52:381–393, 2003.

D. Karlis and X. Pedeli. Flexible bivariate INAR(1) processes using copulas. *Communications in Statistics-Theory and Methods*, 42:723–740, 2013.

D. Karlis and X. Pedeli. Flexible bivariate INAR(1) processes using copulas. *Communications in Statistics-Theory and Methods*, 42:723–740, 2013d.

D. Karlis and E. Xekalaki. Mixed Poisson distributions. *International Statistical Review*, 73:35–58, 2005.

D. Karlis and E. Xekalaki. Mixed Poisson distributions. *International Statistical Review/Revue Internationale de Statistique*, pages 35–58, 2005.

D. Karlis, G.J. Sermaidis, and T. Brijs. Discrete valued time series models for examining weather effects in daily accident counts. *Statistical Modelling*, 2008.

H. Karlsen and D. Tjostheim. Consistent estimates for the near(2) and nlar(2) time series models. *Journal of the Royal Statistical Society-Series B*, 50:313–320, 1988.

H.Y Kim and Y. Park. A non-stationary integer-valued autoregressive model. *Statistical Papers*, 49, 2008.

M. Kirchner. An estimation procedure for the hawkes process. *Quantitative Finance*, 17:571–595, 2017.

R.M. Kirkpatrick and M.C. Neale. Applying multivariate discrete distributions to genetically informative count data. *Behavior Genetics*, 46(2):252–268, 2015.

S.M. Kissler, J.R. Fauver, C. Mack, C.G. Tai, and M.I. Breban. Densely sampled viral trajectories suggest longer duration of acute infection with b.1.1.7 variant relative to non-b.1.1.7 sars-cov-2. *medRxiv 21251535*, 2021.

L.A. Klimko and P.I. Nelson. On conditional least squares estimation for stochastic processes. *Annals of Statistics*, 6:629–642, 1978.

S. Kocherlakota and K. Kocherlakota. *Bivariate Discrete Distributions*, volume 132. CRC Press, illustrated edition, 1992. ISBN 0824787021, 9780824787028.

S. Kocherlakota and K. Kocherlakota. Regression in the bivariate poisson distribution. *Communications in Statistics - Theory and Methods*, 30:815–825, 2001.

C.C. Kokonendji, S. Dossou Gbété, and C.G.B. Demétrio. Some discrete exponential dispersion models: Poisson-tweedie and hinde-demétrio classes. *Statistics and Operations Research Transactions*, 28:201–214, 2004.

C. Lai. Constructions of discrete bivariate distributions. In *Advances in distribution theory, order statistics, and inference*, pages 29–58. Springer, 2006.

D. Lambert. Zero-inflated poisson regression, with an application to defects in manufacturing. *Technometrics*, 34:1–14, 1992.

A. Latour. The multivariate GINAR(p) process. *Advances in Applied Probability*, 29: 228–248, 1997.

A. Latour. Existence and stochastic structure of a non-negative integer-valued autoregressive process. *Journal of Time Series Analysis*, 19:439–455, 1998.

M.L.T. Lee. Properties and applications of the sarmanov family of bivariate distributions. *Communications in Statistics - Theory and Methods*, 25(6):1207–1222, 1996.

T. Liboschik, K. Fokianos, and R. Fried. tscount: An r package for analysis of count time series following generalized linear models. *Journal of Statistical Software*, 82(5):1–51, 2017.

K. Linka, M. Peirlinck, and E. Kuhl. The reproduction number of covid-19 and its correlation with public health interventions. *medRxiv*, 3:1–16, 2020.

Z. Liu and F. Zhu. A new extension of thinning-based integer-valued autoregressive models for count data. *Entropy (Basel, Switzerland)*, 23, 2021.

T. Livio, N. Mamode Khan, M. Bourguignon, and H.S. Bakouch. An inar(1) model with poisson–lindley innovations. *Economics Bulletin*, 38:1505–1513, 2018.

T. Livio, N. Mamode Khan, M. Bourguignon, and H.S. Bakouch. An INAR(1) model with Poisson–Lindley innovations. *Economics Bulletin*, 38(3):1505–1513, 2018.

J. Lorenz. Covid-19 vaccines reduce asymptomatic cases. *Contagion Live*, 2021.

Y. Lu. The predictive distributions of thinning-based count processes. *Scandinavian Journal of Statistics, Wiley, In press*, 1:1–37, 2019.

Yang. Lu. Exact likelihood estimation and probabilistic forecasting in higher-order inar(p) models. *MPRA Paper 83682, Munich:University Library of Munich.*, 2018.

M.E. Mackesy-Amiti, L.J. Ouellet, E.T. Golub, S. Hudson, H. Hagan, and R.S. Garfein. Predictors and correlates of reduced frequency or cessation of injection drug use during a randomized hiv prevention intervention trial. *Addiction*, 106: 601–608, 2011.

N. Mamode Khan, Y. Sunecher, and V. Jowaheer. Modelling a non-stationary BI-NAR(1) Poisson process. *Journal of Statistical Computation and Simulation*, 86: 3106–3126, 2016.

N. Mamode Khan, Y. Sunechar, V. Jowaheer, and M.M. Ristic. Investigating gql-based inferential approaches for non-stationary binar(1)model under different quantum of over-dispersion with application. *Computational Statistics, Springer*, 34:1275–1313, 2018.

N. Mamode Khan, H.O. Cekim, and G. Ozel. The family of the bivariate integer-valued autoregressive process (binar(1)) with poisson-lindley (pl) innovations. *Journal of Statistical Computation and Simulation*, 90:624–637, 2019.

N. Mamode Khan, H. O. Cekim, and G. Ozel. The family of the bivariate integer-valued autoregressive process (BINAR (1)) with Poisson–Lindley (PL) innovations. *Journal of Statistical Computation and Simulation*, 90(4):624–637, 2020.

N. Mamode Khan, A.D. Soobhug, and M. Heenaye-Mamode Khan. Studying the trend of the novel coronavirus series in mauritius and its implications. *PLOS One*, 15(7), 2020a.

N. Mamode Khan, A.D. Soobhug, and M. Heenaye MamodeKhan. Studying the trend of the novel coronavirus series in Mauritius and its implications. *PLOS One*, 15:1–11, 2020b.

N. Mamode Khan, H.S. Bakouch, A.D. Soobhug, and M.G. Scotto. Insights on the trend of the novel coronavirus 2019 series in some small island developing states: A thinning-based modelling approach. *Alexandria Engineering Journal*, 60(2): 2535–2550, 2021.

N. MamodeKhan, A.D. Soobhug, and Z. Jannoo. Modeling road traffic accidents in mauritius using clustered longitudinal com-poisson with gamma random effects. *Communications in Statistics: Case Studies, Data Analysis and Applications*, 7: 113–127, 2021. doi: 10.1080/23737484.2020.1842269.

A. Marshall and I. Olkin. Bivariate distributions generated from polya-eggenberger urn models. *Journal of Multivariate Analysis*, 35:45–65, 1990.

R. Martin. Some results on unilateral arma lattice processes. *Journal for Statistical Planning and Inference*, pages 655–671, 1996.

E.S. McCord and J.H. Ratcliffe. A micro-spatial analysis of the demographic and criminogenic environment of drug markets in philadelphia. *Australian and New Zealand Journal of Criminology*, 40:43–63, 2007.

E. McKenzie. Some simple models for discrete variate time series. *JAWRA Journal of the American Water Resources Association*, 21(4):645–650, 1985.

E. McKenzie. Autoregressive moving-average processes with negative binomial and geometric marginal distributions. *Advances Applied Probability*, 18:679–705, 1986.

E. McKenzie. Some arma models for dependent sequences of poisson counts. *Advances in Applied Probability*, 20:822–835, 1988.

R. Mickens. Difference equations: Theory and application. *Springer*, 1990.

Y. Min and A. Agresti. Random effect models for repeated measures of zero-inflated count data. *Statistical Modelling*, 5(1):1–19, 2005.

M. Mohammadpour, H.S. Bakouch, and M. Shirozhan. Poisson–lindley inar(1) model with applications. *Brazilian Journal of Probability and Statistics*, 32:262–280, 2018.

T.A. Moller, C.H. Weib, H.Y. Kim, and A. Sirchenko. Modeling zero inflation in count data time series with bounded support. *Methodology and Computing in Applied Probability*, 20:589–609, 2017.

T.A. Moller, C.H. Weib, and H.Y. Kim. Modelling counts with state-dependent zero inflation. *Statistical Modelling*, 20:127–147, 2018.

M. Monteiro, M. Scotto, and I. Pereira. A periodic bivariate integer-valued autoregressive model. *Mathematics*, 2015.

D. Morina, J.M. Leyva Moral, and M. Feijoo Cid. Intervention analysis for low-count time series with applications in public health. *Statistical Modelling*, 20:58–70, 2020.

J. Mullahy. Specification and testing of some modified count data models. *Journal of econometrics*, 33(3):341–365, 1986.

J. Murphy. Covid-19 vaccine associated with fewer asymptomatic sars-cov-2 infections. *Pharmacy Times*, 2021.

L. Musango, L. Veerapa Mangroo, Z. Joomaye, A. Ghurbhurrun, V. Vythelingam, and E. Paul. Key success factors of mauritius in the fight against covid-19. *BMJ Global Health*, 6, 2021.

M.Yang, G.Zamba, and J.Cavanaugh. Zim: Zero-inflated models (zim) for count time series with excess zeros, r package version 1.1.0. 2018. URL `https://cran.r-project.org/package=ZIM`.

A.S. Nastic and H.S. Bakouch. A combined geometric inar(p) model based on negative binomial thinning. *Mathematical and Computer Modelling*, 55:1665–1672, 2012.

A.S. Nastic, M.M. Ristic, and P.M. Popovic. Estimation in a bivariate integer-valued autoregressive process. *Communication in Statistics-Theory and Methods*, 45: 5660–5678, 2016.

A.S. Nastic, M. M. Risti, and A. D. Janji. A mixed thinning based geometric inar(1) model. *Filomat*, 31:4009–4022, 2017.

J.M. Neuhaus, C.E. McCulloch, and R. Boylan. Estimation of covariate effects in generalized linear mixed models with a misspecified distribution of random intercepts and slopes. *Journal of Multivariate Analysis*, 32:2419–2429, 2013.

C.M. Ng, S. Onga, and H.M. Srivastava. A class of bivariate negative binomial distributions with different index parameters in the marginals. *Applied Mathematics and Computation*, 217:3069–3087, 2010.

K. Nytholm. Inferring the private information content of trades: A regime-switching approach. *Journal of Applied Econometrics*, 18(4):457–470, 2003.

X. Pedeli and D. Karlis. A bivariate INAR(1) process with application. *Statistical Modelling: An International Journal*, 11(4):325–349, 2011.

X. Pedeli and D. Karlis. A bivariate INAR (1) process with application. *Statistical Modelling*, 11(4):325–349, 2011.

X. Pedeli and D. Karlis. On composite likelihood estimation of a multivariate inar(1) model. *Journal of Time Series Analysis*, 34:206–220, 2013aa.

X. Pedeli and D. Karlis. Some properties of multivariate INAR(1) processes. *Computational Statistics and Data Analysis*, 67:213–225, 2013ab.

X. Pedeli and D. Karlis. On estimation of the bivariate poisson INAR process. *Communications in Statisitcs-Theory and Methods*, 35:514–533, 2013b.

X. Pedeli and D. Karlis. Some properties of multivariate inar(1) processes. *Computational Statistics and Data Analysis*, 67:213–225, 2013ca.

X. Pedeli and D. Karlis. On composite likelihood estimation of a multivariate INAR(1) model. *Journal of Time Series Analysis*, 34:206–220, 2013cb.

X. Pedeli, A. Davidson, and K. Fokianos. Likelihood estimation for the inar(p) model by saddlepoint approximation. *Journal of the American Statistical Association*, 110:1229–1238, 2015a.

X. Pedeli, Anthony C. Davison, and Fokianos. Konstantinos. Likelihood estimation for the inar(p) model by saddlepoint approximation. *Journal of the American Statistical Association*, 110:1229–1238., 2015b.

S.E. Perumean-Chaney, C. Morgan, D. McDowall, and I. Aban. Zero-inflated and overdispersed: what's one to do. *Journal of Statistical Computation and Simulation*, 83:1671–1683, 2012.

D. Pickard. Unilateral markov fields. *Advances in Applied Probability*, 12:655–671, 1980.

A. Pirdavani, T. Brijs, and T. Bellemans. Spatial analysis of fatal and injury crashes in flanders, belgium; application of geographically weighted regression technique. *Conference: 92nd Transportation Research Board Conference*, 2013.

P. M. Popovic, A.S. Nastic, and M.M. Ristic. Residuals analysis with bivariate inar(1) model. *REVSTAT – Statistical Journal*, 16:349–363, 2018a.

P.M. Popovic, M.M. Ristic, and A.S. Nastic. A geometric bivariate time series with different marginal parameters. *Statistical Papers*, 57:731–753, 2018b.

A.M.M.S. Quoreshi. Bivariate time series modeling of financial count data. *Communication in Statistics-Theory and Methods*, 35:1343–1358, 2006.

S. Quoreshi, R. Uddin, and N. Mamode Khan. Quasi-maximum likelihood estimation for long memory stock transaction data—under conditional heteroskedasticity framework. *Journal of Risk and Financial Management*, 12:1–13, 2020.

O.O. Rasaki. Integer-valued time series model via generalized linear models technique of estimation. *International Annals of Science*, 4:35–43, 2018.

M. M. Ristic and B. Popovic. A new bivariate binomial time series model. *Markov Processes and Related Fields*, 25(2):1–26, 2019.

M.M. Ristic, H.S. Bakouch, and A.S. Nastic. A new geometric first-order integer-valued autoregressive (NGINAR(1)) process. *Journal of Statistical Planning and Inference*, 136:2218–2226, 2009.

M.M. Ristic, A.S. Nastic, K. Jayakumar, and H.S. Bakouch. A bivariate INAR(1) time series model with geometric marginals. *Applied Mathematical Letters*, 25: 481–485, 2012.

H. Ritchie, E. Ortiz-Ospina, D. Beltekian, E. Mathieu, J. Hasell, B. Macdonald, C. Giattino, C. Appel, L. Rodés-Guirao, and M. Roser. Our world in data : Covid-19: Stringency index. 2021. URL https://ourworldindata.org/covid-stringency-index.

C.E. Rose, S.W. Martin, K.A. Wannemuehler, and B.D. Plikaytis. On the use of zero-inflated and hurdle models for modeling vaccine adverse event count data. *Journal of biopharmaceutical statistics*, 16(4):463–481, 2006.

A. Safari, R. MacKay Altman, and B. Leroux. Parameter-driven models for time series of count data. *The Canadian Journal of Statistics*, 1:1–29, 2011.

Z. Sajjadnia, M. Sharafi, N. Mamode Khan, and A.D. Soobhug. The bivariate INAR(1) model with paired Poisson-weighted exponential distributions. *(Communicated)*, 2021.

José María Sarabia Alegría, Emilio Gómez Déniz, et al. Construction of multivariate distributions: a review of some recent results. 2008.

O. Sarmanov. Generalized normal correlation and two-dimensional frechet classes. *Doklady Soviet Mathematics Tom*, 168:596–599, 1966.

O. V. Sarmanov. Generalized normal correlation and two-dimensional fréchet classes. In *Doklady Akademii Nauk*, volume 168, pages 32–35. Russian Academy of Sciences, 1966.

S. Schweer and C. Weiß. Compound poisson inar(1) processes: Stochastic properties and testing for overdispersion. *Computational Statistics and Data Analysis*, 77: 267–284, 2014.

M. Scotto, C. Weiss, M. Silva, and I. Pereira. Bivariate binomial autoregressive models. *Journal of Multivariate Analysis*, 125:233–251, 2014a.

M. Scotto, C. Weiß, M.E. Silva, and I. Pereira. Bivariate binomial autoregressive models. *Journal of Multivariate Analysis*, 125:233–251, 2014b.

M. G. Scotto, C. H. Weiß, and S. Gouveia. Thinning-based models in the analysis of integer-valued time series: A review. *Statistical Modelling*, 15:590 – 618, 2015.

E. Sekyere, N. Bohler Muller, C. Hongoro, and Dr. M. Makoae. Africa program occasional paper - the impact of covid-19 in south africa. 2017.

K.F. Sellers, G. Shmueli, and S. Borle. The com-poisson model for count data: a survey of methods and applications. *Applied Stochastic Models in Business and Industry*, 28:104–116, 2011.

K.F. Sellers, D.S. Morris, and N. Balakrishnan. Bivariate conway–maxwell–poisson distribution: Formulation, properties, and inference. *Journal of Multivariate Analysis*, 150:152–168, 2016.

G. Shafabakhsh, A. Famili, and M. Bahadori. Gis-based spatial analysis of urban traffic accidents: Case study in mashhad, iran. *Journal of Traffic and Transportation Engineering (English Edition)*, 4(3):290–299, 2017.

A.S.V. Shah, C. Gribben, J. Bishop, P. Hanlon, D. Caldwell, R. Wood, M. Reid, J. McMenamin, D. Goldberg, D. Stockton, S. Hutchinson, C. Robertson, P.M. McKeigue, H.M. Colhoun, and D.A. McAllister. Effect of vaccination on transmission of covid-19: an observational study in healthcare workers and their households. *medRxiv*, 2021. doi: 10.1101/2021.03.11.21253275.

M. Sharafi, Z. Sajjadnia, and A. Zamani. A first order integer-valued autoregressive process with zero modified poisson-lindley distributed innovations. *Communications in Statistics - Simulation and Computation*, 2021.

G. Shmueli, T.P. Minka, J.B. Kadane, S. Borle, and P. Boatwright. A useful distribution for fitting discrete data: revival of the Conway - Maxwell - Poisson distribution. *Applied Statistics*, 54:127–142, 2005.

I. Silva, M. E. Silva, I. Pereira, and N. Silva. Replicated inar(1) processes. *Methodology and Computing in Applied Probability*, 7:517–542, 2005.

A.D. Soobhug, H. Jowaheer, N. Mamode Khan, N. Reetoo, K. Meethoo Badulla, L. Musango, C.C. Kokonendji, A. Chutoo, and N. Aries. Re-analyzing the sars-cov-2 series using an extended integer-valued time series models: A situational assessment of the covid-19 in mauritius. *PLOS One*, 2022.

F.W. Steutel and K. Van Harn. Discrete analogues of self-decomposability and stability. *The Annals of Probability*, 7:893–899, 1979.

Y. Sunechar, N. Mamode Khan, and V. Jowaheer. A gql estimation approach for analysing non-stationary over-dispersed binar(1) time series. *Journal of Statistical Computation and Simulation*, 87:1911–1924, 2017.

Y. Sunechar, N. Mamode Khan, and V. Jowaheer. Estimation methods for a flexible INAR(1) COM-Poisson time series model. *Journal of Applied Mathematics, Statistics and Informatics*, 14:57–82, 2018a.

Y. Sunechar, N. Mamode Khan, M.M. Ristic, and V. Jowaheer. Binar(1) negative binomial model for bivariate non-stationary time series with different over-dispersion indices. *Statistical Methods and Applications, Springer*, 28:625–653, 2018b.

M. Tang and Y. Wang. Asymptotic behavior of random coefficient inar model under random environment defined by difference equation. *Advance in Difference Equations*, 99, 2014.

K. Tawiah, W.A. Iddrisu, and K. A. Asosega. Zero-inflated time series modelling of covid-19 deaths in ghana. *Journal of Environmental and Public Health*, 2021:1–9, 2021.

Mei-Ling Ting Lee. Properties and applications of the sarmanov family of bivariate distributions. *Communications in Statistics-Theory and Methods*, 25(6):1207–1222, 1996.

D. Tjostheim. Estimation in non-linear time series models. *Stochastic Processes and their Applications*, 21:251–273, 1986.

J.W. Tukey. Moments of random group size distributions. *Annals of Mathematical Statistics*, 20:523–539, 1949.

K.F. Turkman, M.G. Scotto, and P. de Zea Bermudez. Non-linear time series: extreme events and integer value problems. *Switzerland: Springer-Verlag.*, pages 1–255, 2014.

R. Tyagi, L. K. Dwivedi, and A. Sanzgiri. Estimation of effective reproduction numbers for covid-19 using real-time bayesian method for india and its states. *Research Square*, 1, 2020. doi: 10.21203/rs.3.rs-45937/v1.

L. Wang, H. Aldirawi, and J. Yang. Identifying zero-inflated distributions with a new r package izid. *Journal of Communication in Information and Systems*, 20:23–44, 2020.

A.K. Weaver, J.R. Head, C.F. Gould, E.J. Carlton, and J.V. Remais. Environmental factors influencing covid-19 incidence and severity. *Annual Review of Public Health*, 43:271–291, 2012.

C. Weiß. A poisson inar(1) model with serially dependent innovations. *Metrika*, 78: 829–851, 2015.

C. H. Weiß. Thinning operations for modeling time series of counts - a survey. *Springer-Verlag*, 92:319–341, 2008.

C. H. Weiß, F. Zhu, and A. Hoshiya. Softplus ingarch models. *Statistica Sinica*, pages 1–41, 2021.

Christian. H. Weiß. *An Introduction to Discrete-Valued Time Series*. Wiley, Chichester, 2018.

R. Xu, H. Rahmandad, M. Gupta, C. Digennaro, N. Ghaffarzadegan, H. Amini, and M.S. Jalali. Weather conditions and covid-19 transmission: Estimates and projections. *medRxiv*, 2020.

F. Ye, T.P. Garcia, M. Pourahmadi, and D. Lord. Extension of a negative binomial garch model: Analyzing the effects of gasoline price and vmt on dui fatal crashes in texas. *91st Annual Meeting of the Transportation Research Board*, pages 1–17, 2011.

K.C.H. Yip and K.K.W. Yau. On modeling claim frequency data in general insurance with extra zeros. *Insur.: Math. Econ.*, 39:153–163, 2005.

M. Yu, D. Wang, K. Yang, and Y. Liu. Bivariate first-order random coefficient integer-valued autoregressive processes. *Journal of Statistical Planning and Inference*, 204:153–176, 2020.

X. Yu, M. Baron, and P.K. Choudhary. Change-point detection in binomial thinning processes, with applications in epidemiology. *Sequential Analysis*, 32:350–367, 2013.

S. Zhang, Yuan. Liao, and Wei. Ning. Asymptotic properties of quasi-maximum likelihood estimates in generalized linear models. *Communications in Statistics—Theory and Methods*, 40:4417–4430, 2010.

S. Zhang, , D. Wang, and X. Fan. A negative binomial thinning-based bivariate inar(1) process. *Statistica Neerlandica*, 74:517–537, 2020.

F. Zhu. A negative binomial integer-valued garch model. *Wiley Online Library*, pages 1–14, 2009.

F. Zhu. Modeling overdispersed or underdispersed count data with generalized poisson integer-valued garch models. *Journal of Mathematical Analysis and Applications*, 389:58–71, 2012a.

F. Zhu. Zero-inflated poisson and negative binomial integer-valued garch models. *Journal of Statistical Planning and Inference*, 142:826–839, 2012b.

F. Zhu. Modeling time series of counts with com-poisson ingarch models. *Journal of Mathematical and Computer Modelling, Elsevier*, 56:1–13, 2012c.

F. Zhu and H. Joe. Modelling count data time series with markov processes based on binomial thinning. *Computational Statistics and Data Analysis*, 27:725–738, 2006.

F. Zhu and Q. Li. Moment and bayesian estimation of parameters in the ingarch (1,1) model (in chinese). *Journal of Jilin University (Science Edition)*, 47:899–902, 2009.

R. Zhu and H. Joe. Count data time series models based on expectation thinning. *Stochastic Models*, 26:431–62, 2010a.

R. Zhu and H. Joe. Negative binomial time series models based on expectation thinning operators. *Journal of Statistical Planning and Inference*, 140:1874–88, 2010b.

A.F. Zuur, E.N. Ieno, N.J. Walker, A.A. Saveliev, and G.M. Smith. Mixed effects models and extensions in ecology with r. *New York, NY: Springer Science and Business Media.*, 2009.

Index

Note: Pages in **bold** and *italics* refer to tables and figures, respectively.

For Product Safety Concerns and Information please contact our EU
representative GPSR@taylorandfrancis.com
Taylor & Francis Verlag GmbH, Kaufingerstraße 24, 80331 München, Germany